Fluvial Geomorphology
of Australia

Fluvial Geomorphology of Australia

Edited by
Robin F. Warner
Department of Geography
University of Sydney, Australia

ACADEMIC PRESS
Harcourt Brace Jovanovich, Publishers
Sydney San Diego New York Berkeley
Boston London Tokyo Toronto

ACADEMIC PRESS AUSTRALIA
30–52 Smidmore Street, Marrickville, NSW 2204

United States Edition published by
ACADEMIC PRESS INC.
1250 Sixth Avenue
San Diego, California 92101–4311

United Kingdom Edition published by
ACADEMIC PRESS, INC. (LONDON) LTD.
24/28 Oval Road, London NW1 7DX

Printed in Australia

National Library of Australia Cataloguing-in-Publication Data

Fluvial geomorphology of Australia.

Includes index.
ISBN 0 12 735660 6.

1. Geomorphology — Australia.   2. Sedimentation and
deposition — Australia.   I. Warner, Robin F.

551.4'0994

Library of Congress Catalog Card Number: 88-071399

Contributors

Numbers in parentheses indicate pages on which the authors' contributions begin.

G. Allan (177), CSIRO Division of Wildlife and Ecology, Alice Springs, NT.

V. R. Baker (177), Department of Geosciences, University of Arizona, Tucson, Arizona, USA.

Juliet Bird (343), Environmental Science Centre, Melbourne College of Advanced Education, Carlton, Victoria.

Mike Bonell (41), Department of Geography, James Cook University of North Queensland, Townsville, Qld.

Bryan L. Campbell (87), Radioisotope Research Section, Nuclear Science & Technology Organisation, Lucas Heights, NSW.

R. F. Cull (303), Alligator Rivers Region Research Institute, Jabiru East, NT.

Kate Duggan (303), NTTIAD Project, Kupang, Timor, Indonesia.

T. J. East (303), Alligator Rivers Region Research Institute, Jabiru East, NT.

Gregory L. Elliott (87), Soil Conservation Service of NSW Research Centre, Gunnedah, NSW.

Wayne D. Erskine (201, 223), School of Geography, University of New South Wales, Kensington, NSW.

B. L. Finlayson (17), Department of Geography, University of Melbourne, Parkville, Victoria.

Robert J. Loughran (87), Department of Geography, University of Newcastle, NSW.

T. A. McMahon (17), Department of Civil and Agricultural Engineering, University of Melbourne, Parkville, Victoria.

A. S. Murray (303), CSIRO Division of Water Resources Research, Canberra, ACT.

Gerald C. Nanson (151, 201), Department of Geography, University of Wollongong, NSW.

R. J. Neller (323), Department of Geography, School of Australian Environmental Studies, Griffith University, Brisbane, Qld.

L. J. Olive (69, 283), Department of Geography and Oceanography, Australian Defence Force Academy, Campbell, ACT.

K. J. Page (267), School of Applied Science, Riverina-Murray Institute of Higher Education, Wagga Wagga, NSW.

Geoff Pickup (105, 177), CSIRO Division of Wildlife and Ecology Station, Alice Springs, NT.

David M. Price (151), Department of Geography, University of Wollongong, NSW.

W. A. Rieger (69, 283), Department of Geography and Oceanography, Australian Defence Force Academy, Campbell, ACT.

S. J. Riley (245), School of Earth Sciences, Macquarie University, Sydney, NSW.

Brian R. Rust (151), Department of Geology, University of Ottawa, Canada.

Robin F. Warner (1, 223, 343), Department of Geography, University of Sydney, NSW.

Karl-Heinz Wyrwoll (129), Department of Geography, University of Western Australia, Nedlands, WA.

Robert W. Young (151), Department of Geography, University of Wollongong, NSW.

Preface

The idea for this book occurred as long ago as 1983 when Alistair Pitty invited me to contribute a chapter on Australian fluvial geomorphology to his book, *Themes in Geomorphology* (1984).

The sections of the essay which appeared seemed to be ideal for expanding into a book on fluvial geomorphology in Australia, especially as I have had 25 years experience of working in down-under rivers. So a book seemed like a good idea but, while publishers were enthusiastic about the content, they were doubtful about the market. It is certainly true that there are many good fluvial texts already, but Australia is different and generalisations made, based on North American and European work, do not seem to hold so readily here. This is often confusing for students and researchers alike. These differences have been addressed in many of the essays which follow.

Ideas for a volume of essays were discussed with several colleagues; notably Gerald Nanson and Bob Young at Wollongong, Geoff Pickup at Alice Springs, and Laurie Olive and Bill Rieger at the Australian Defence Force Academy in Canberra. Such discussions were profitable in setting up the book as well as in resulting contributions from those sources.

This volume should fill a significant gap for local students and researchers in terms of the differences already referred to and as background for their own studies and work. It should also be useful for overseas colleagues who perhaps may not believe that conditions are different — or who may be intrigued enough to come down under to help with the work still to be done. Other locals, like friendly and enlightened engineers, environmental specialists and perhaps more perceptive planners, may find much to interest them. There is much to know and to find out about our rivers and how they affect our lives.

Finally, two other events seem to help make this book timely. This is Australia's bicentennial year and fluvial geomorphologists might celebrate with a book of this kind. This is also the year when the International Geographical Union meets in Sydney. Those of our overseas colleagues who have the good fortune to visit us may find this collection both interesting and relevant.

Robin F. Warner

Acknowledgements

This book could not have been completed without the help of many people. The editor is grateful to Wayne Erskine for his considerable assistance in reviewing some essays and for final proofing. Other colleagues who helped in the review stage include: Geoff Pickup, Bill Rieger, Laurie Olive, Gerald Nanson, Mike Melville and Brian Zaitlin. This help is gratefully acknowledged.

Chapters were typed, often from very battered and some very late manuscripts, in a very friendly and patient manner by Janette Brennan. John Roberts, the cartographer, not only produced many figures but did near-impossible reduction jobs with drawings of variable quality and size. Their efforts were much appreciated as was their tolerance.

The editor also acknowledges the efforts of his friends and colleagues who took part in this venture. They produced the essays which demonstrate many of the current interests in Australian fluvial geomorphology. He was especially pleased with those authors who delivered clean manuscripts on time, but the efforts of all are much appreciated because several contributors were known to have incredible backlogs of work and commitments.

Indeed some of those invited could not produce because of such pressures. It is to be hoped that their contributions will be found in the next volume which, given the rate of current developments in the subject, should not be too long in appearing.

Contents

1

Fluvial Geomorphology in Australia: Some Perspectives

Robin F. Warner

Department of Geography
University of Sydney
Sydney,   N.S.W.

I. INTRODUCTION

This is the first collection of essays on Australian
fluvial geomorphology. In it an attempt is made to show
something of the geomorphology of rivers and related
phenomena in this island continent, as well as some of the
contemporary diversity of interests in this subject. The
approach is neither systematic nor well balanced. It
cannot be the former because of gaps in the presentation.
It is not the latter because most attention has always been
focused in the south east, where most of the people live,
where most of the research institutes are located and
where most of the geomorphologists pursue their interests.

However, the range of topics and their geographic spread
show many of the present-day major thrusts being made by
colleagues. Australia is a very large and, for the most
part, flat dry land mass. With about 7.7 mill km^2 landsurface area and probably fewer than 50 geomorphologists,
there is a very large backyard for each individual. More
realistically and, because we tend to step on each other's
toes, vast areas will never be investigated. Since about
two thirds is semiarid or arid, students of fluvial
processes tend not to find much action away from coastal
fringes, but many of the arid areas have landform
assemblages dominated by fluvial forms.

Only short peripheral rivers flow directly into the sea
with the notable exception of the 1 mill km^2 basin of the

1

FLUVIAL GEOMORPHOLOGY OF AUSTRALIA
ISBN   0 12 735660 6

Murray Darling. To the north of this, the larger, sometimes integrated Lake Eyre Basin has no external outlet.

The main aims of this brief introductory essay are to:
(i) describe the land platform and its evolution; (ii) review the main areas of research in fluvial geomorphology, particularly leading up to this book; (iii) assess individual and group impacts on the development of the subject, and (iv) outline the following collection of essays.

II. THE LAND PLATFORM AND ITS EVOLUTION

Since much of this book is concerned with contemporary processes and forms, this section attempts to highlight briefly features of the land mass, its early evolution, its Quaternary evolution and the present.

A. *The Land Mass*

The land surface is made up of three major provinces: the western plateaux, the central lowlands and the eastern highlands (Jennings and Mabbutt, 1977). More than half is dominated by the ancient, planed-down, stable low-lying plateaux of west and central Australia. The eastern highlands are composed of mainly Lower Palaeozoic sediments which have been much altered and intruded by granites. They have been uplifted and planed down on several occasions. Mesozoic basins interrupt the north-south continuity and in many places they have been capped by Tertiary basalts. The central lowlands have been sites for shallow seas and have thus been sediment sinks for material derived from west and eastern blocks. Strata are mainly young and not greatly disturbed. Consequently they form the low plains of the Gulf of Carpentaria, the Lake Eyre and the Murray-Darling basins. In many parts they are still aggradational surfaces for subaerial wastes.

B. *Its Early Evolution*

This platform was for a long time part of the larger, more contiguous group of continents called Gondwanaland (Ollier, 1986). The Australian Plate has broken away progressively from bordering plates since the Jurassic. It has also drifted northwards into warmer latitudes, as is evident from palaeomagnetic studies (Pillans, 1983).

New margins and therefore new baselevels for rivers were formed at different times with these ruptures. Some old river systems were truncated by such drifting. In the north west, the break with India and Africa began early, while the Perth Basin (Fig. 1, for location) to the south was exposed later (Ollier, 1986). Sea-floor spreading opened up the Tasman Sea in the south east between 60 and 80 mya and the Coral Sea to the north east 56-62 mya, while the break with Antarctica to the south was later (Ollier, 1986). Since the break ups, there has been some erosion and some sedimentation along the margins but low rates of denudation have prevailed from a land mass where nearly 40% is lower than 200 m (Jennings and Mabbutt, 1986).

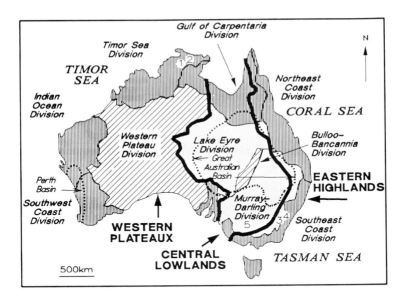

FIGURE 1. Map showing locations mentioned in the text. Numbers on the map refer to the following locations:

1. South Alligator River
2. East Alligator River
3. Southeast Highlands (Kosciusko Area)
4. Lake George
5. Riverine Plain of the Murrumbidgee River.

With these detachments the land mass took on the shape now known. While there have been no subsequent mountain-building episodes, broad tectonic warping has elevated many of the planed-down surfaces to form uplands. It has also been responsible for creating the low divides in the central lowlands (Ollier, 1986).

With parts of the continent being very old (some parts have been land since the Palaeozoic), with other parts being affected by warping and still others subject to shallow-sea transgressions, it is no wonder, as Ollier (1986) points out, that the ages of the drainage pattern vary widely. In Western Australia, fossil drainage lines dating back to the Lower Palaeozoic still characterise the plateaux surfaces. To the north and centre Permian glaciations and Cretaceous seas affected large areas to form remodelled and new surfaces respectively. On their lower margins older rivers flowed out over newly emerged surfaces. Surface drainage in the Great Artesian Basin to the east was consequent to uplift following Mesozoic deposition. Farther south, Eocene origins have been suggested for the Murray and Darling basins which are separated by the Cobar Ridge (Ollier, 1986).

The drainage areas have been subjected to various regimes imposed by different climates as the land mass drifted northwards relative to the shifting south pole (Pillans, 1983). Differential uplifts following the post-Gondwana rupture have created highs and lows on deeply weathered surfaces. These have been researched in many parts (see books edited by Jennings and Mabbutt, 1967; Davies and Williams, 1978; Langford-Smith, 1980). Apparent unity of the eastern highlands prompted Andrews (1910) to suggest that there had been an extensive surface which had been uplifted at the end of the Tertiary in what he called the Kosciusko Uplift. This notion, which was perpetuated by King (1962) and others, was not in accord with geological evidence (Bishop, 1982). Reinterpretation of early ideas on Tertiary rivers in Eastern Australia has been possible through the dating of widespread basalts, which fossilized deposits on pre-existing surfaces, the determination of former stream directions and the use of other recently developed geological techniques (see reviews: Bishop, 1982; Ollier, 1986; Young, 1981). It now appears that divides have been stable and that any uplifts are much older than originally thought.

C. *Quaternary Evolution*

This period of only 2.5 my has been characterised by alternating climates and sea-level shifts associated with glacials and interglacials. These have caused regime variations which were not necessarily great because low altitudes and generally low latitudes meant that only small parts of Tasmania and the south-eastern highlands were glaciated. Rises and falls of sea level only affected the base levels of small peripheral rivers, rejuvenating them in low sea-level stands and drowning them in high. The combined impacts of climatic and sea-level changes did cause terracing which up until recently received only limited treatment (review: Warner, 1972; Walker and Coventry, 1976).

Inland valleys were less constricted than near the coast. Alluvial, lacustrine and aeolian stratigraphies provide a wealth of evidence for complex Quaternary evolution. This has been reviewed in detail by Bowler (1986). His careful interpretations have superceded the ideas of Butler (1967), Schumm (1968) and others. Much of this work was concentrated on the Murrumbidgee and Murray but similar evolution patterns are now becoming evident from the Darling and the Channel Country (Nanson *et al*, this volume).

Large channel dimensions were created by bank failure associated with increased pore-water pressure combined with higher water tables (Bowler, 1986). Sandier channels and source-bordering dunes were promoted in cold periods by the absence of vegetation and greater aeolian activity. With the return of vegetation about 12 ka, channels became more sinuous, thereby decreasing slope and bed-load movement. More recent high flows have been accommodated by avulsion and flood-plain storages, rather than by in-channel changes.

Lakes as sediment sinks have been useful in working out long-term evolution, as at Lake George (Singh *et al*, 1981). The reinterpretation of some of the coastal terrace sequences is now under way with recent work in the Hawkesbury Nepean (Nanson and Young, 1987; Nanson *et al*, 1987), showing that some of the terraces are much older than previously thought. This is in line with inland findings on the Tallygaroopna sequence at 36 to 50 ka (Bowler, 1986).

The Quaternary, Bowler (1986) argued, was not only characterised by climatic changes but their impacts on water balances, water tables and aeolian activity. These are important in a land where the role of glaciation was small.

D. The Present

    Australia is presently a dry continent with a mean annual
precipitation of only 420 mm and an average runoff depth
of only 45 mm (Warner, 1986). It is little wonder that
denudation rates are low in undisturbed areas. However,
averages are fairly meaningless in a land where the dead
heart can be revitalised overnight by megafloods (Pickup *et
al*, this volume) and where in north-east Queensland 4 m of
annual rainfall are common (Bonell, this volume).
    The present fluvial landscape is not one of a simple,
dynamic balance between morphology and processes. In many
parts of the south east regimes have alternated between
flood- and drought-dominated modes, with attendant
instability and lagged adjustments (Warner, 1987a; 1987b;
Erskine and Warner, this volume). Superimposed on these
spatially variable components of channel morphology involving
the five long-profile zones (Pickup, 1986) are the impacts
of catastrophic modifications to alluvial morphology (Nanson,
1986). These regime shifts are related to secular variations
in precipitation (Gentilli, 1971; Cornish, 1977; Riley, 1981)
and are referred to several times in this volume (Erskine
and Warner; Riley; Page; and Nanson and Erskine).
    Energy provided by moving water removes material from
source areas, transmits it intermittently through a network
of temporary alluvial stores and into sinks or mud basins
(Pickup, 1986; Schumm, 1977). Continuity demands that
adjustments occur throughout the system, but spatial
variations mean that the links between discharge and channel
morphology are more tenuous than previously thought by Dury
et al (1963) and Woodyer (1968).
    Many of the essays in this volume address present-day
problems in fluvial geomorphology and there is little need
to elaborate further here.

III.   MAIN AREAS OF RESEARCH

    Fluvial geomorphology covers numerous areas of research
including: processes and morphology, their interrelation-
ships, past and present alluviation, river evolution and
valley-fill chronologies of valleys and a number of more
specialist topics. Consideration of these has changed with
time, with the development of new techniques for interpreting
evidence, increased data availability and with those
individuals who have influenced the directions of change

and development. Consequently it is possible to comment
briefly on the subject and how it has changed over time.
Three periods seem appropriate: pre-1945, 1945 to 1979 and
1980 to the present.

A. *The Early Days*

In the late 19th Century, geologists devoted some
attention to landscape evolution. There was much to
discover and map in this vast land by researchers like David
(1877). He was concerned with fossil gravels in tin leads,
as were Browne and Raggatt (review: 1935). The visit of
David's student, Andrews, to ₁the United States allowed rapid
application of some of Gilbert's ideas, but more
particularly those of W.M. Davis, in Australia (Andrews,
1903 and 1910). Another student of David, T.G. Taylor
become the Commonwealth Bureau of Meteorology's first and
only physiographer (1911). His geomorphological interests
were evolution of drainage, stream capture and the role of
tectonic relief. After opening the first Geography
Department at the University of Sydney in 1920, he developed
much wider interests before leaving for Canada in 1928. This
was to remain the only Geography Department in Australia
until 1945.
Surfaces and drainage evolution in coastal New South
Wales were considered by Craft, a student of Taylor's, in
17 papers in the 1930s (see Browne, 1975 for list).
Physiographic contributions for New England and Victoria were
made by Voisey (1942) and Hills (1939) respectively (see
Joyce, 1987 for list of Hills' 114 works).
There was little on process geomorphology, although
Josephson (1885) and Hall (1927) presented detailed studies
of flooding on the Hawkesbury. In more recent times, many
notable engineering studies have been uncovered (eg, Maiden,
1903) but it was a long time before geomorphologists
exploited such information (Erskine and Warner; Warner and
Bird, this volume).

B. *1946 to 1979*

In the period after the Second World War, there were
changes in subject emphasis and personnel. Much more of
the land surface was "discovered", described and analysed
by CSIRO land resource research teams. This was helped
considerably with the more readily available vertical air
photographs in areas for which there were few if any maps.

Also in the early 1950s, British-trained geomorphologists like Jennings, Mabbutt and Twidale arrived, the latter two initially with CSIRO. Under these teams, remote landscapes were classified into land systems based on geology, geomorphology, soils, climate and vegetation. The emphasis perhaps was still on regional landscapes and their evolution, rather than on rivers and their processes.

A brief influx of migrant geomorphologists followed in the early 1960s, but even then there was still a bias towards landscape evolution caused by the dominance of Davis' ideas and denudation chronology in the British system. There was something of a reaction to this particularly on the Riverine Plains by mainly local soil scientists (Butler, 1958; 1961), hydrologists (Pels, 1964; 1966), geographers (Langford-Smith, 1960) and geologists (Schumm, 1968). Also in the early 1960s, quantitative geomorphology, underfit streams and other topics were introduced with the arrival of George Dury (1964, 1970, 1976 etc.). In his graduate school, there were additional research interests on terraces (Hickin, 1970) point dunes (Hickin, 1969), meanders (Young, 1970), non-tectonic evolution (Young, 1978), channel patterns of western rivers (Riley, 1977), drainage networks (Abrahams, 1972), and so on.

Two papers help to indicate major thrusts and main developments of this time: Spate and Jennings (1972) on Australian geography from 1951-1971, and Jeans and Davies (1984) from 1972 to 1982. In the former, 95 papers were listed by Jennings. Of these 32 were broadly fluvial in context, with 14 on evolution and Quaternary stratigraphy, 5 on quantitative, 5 on sediment yield, processes and water quality, 4 on meanders and 4 on floods and bankfull phenomena. More than half were published between 1965 and 1969. Over the next 11 years output doubled the earlier rate, with 9 on the Quaternary, 6 on evolution, 6 on processes and hydrology, 4 on water quality, and 3 each on quantitative topics and channel change/alluvial morphology. There were needless to say many omissions from these reviews but they did reveal the increase in output and changes in subject matter.

The period was also reviewed to some extent in important edited books (Jennings and Mabbutt, 1967; Jeans, 1977; Davies and Williams, 1978; Langford-Smith, 1980). These were particularly strong in their coverage of landscape evolution but there was little on fluvial geomorphology specifically. Individual books on humid landforms (Douglas, 1977), desert landforms (Mabbutt, 1977) and more general geomorphology (Twidale, 1976) also did much to summarise developments in this and related subjects.

C. The Present

Fluvial geomorphology has blossomed even more in the 1980s, in spite of little growth in the university system and often tighter funding. There are more fluvial specialists, some of whom have university positions but others are in government agencies and private companies.

There is now a much stronger association with engineers than geologists and a lot of the work is applied to some extent (Warner and Bird, this volume). There is probably less emphasis on early evolution with some notable exceptions (Ollier, 1986; Bishop, 1982; Young, 1981). However, Quaternary stratigraphies are still attracting much attention: on the plains (Bowler, 1986; Nanson *et al*, this volume); in the north (Woodroffe *et al*, 1987); and in the west (Wyrwoll, this volume). River terraces have recently been reinvestigated (Atkinson, 1987; Nanson and Young, 1987; Nanson *et al*, 1987). Many of the other areas of contemporary interest in this subject are addressed in the following essays. The literature appraisals in them will add considerably to this brief statement. Increased interest in the scientific study of rivers and the practical values of such works were the main reasons for assembling this collection.

IV. THE IMPACTS OF INDIVIDUALS AND GROUPS

Fluvial geomorphology had its roots in the last century with geologists, some of whom were soon influenced by American pioneers like Davis. Sir Thomas Edgeworth David, a Welshman by birth, had impacts as a geological surveyor and later as an academic. These were continued by his students: T.G. Taylor, Browne, Andrews, Woolnough, Hills and others. Their influence on the present is largely indirect from their writings. Direct links with the present are difficult to find. Andrews stayed in government service. Taylor went off to Canada and his student, Craft, had to return to school teaching in the Depression when his Linnean Society fellowship terminated.

Since there was only one geography school at the University of Sydney until 1945, the second wave of those influencing directions and content were either post-World War II imports or homespun products like Langford-Smith.

Some of the latter's interests were in fluvial geomorphology (1960), although he became more concerned with arid and coastal topics. Notable imports were Jennings, Mabbutt and Dury.

Jennings' work in karst and coast is well known but his great love of geomorphology, his help and thought-provoking comments were inspirational. He was much respected for over 30 years (Jeans and Davies, 1984; Davies and Williams, 1978; Bowler, 1986). Great insight into his love affair with this country and its landscapes was provided in one of his last papers - the Introduction in Jeans' *Australia - a Geography* (1986).

Dury spent most of the 1960s at Sydney before leaving for Wisconsin in 1970. His graduates are well known in Australia, North America, Africa and England. His extension work in schools and great enthusiasm for the subject did much to popularise fluvial geomorphology.

Douglas, a graduate of Jennings, held the chair at New England in the 1970s before going to Manchester. His contributions to denudation rates and urban geomorphology are well known (1973; 1983). He did much to foster small catchment studies and urban research (Neller, this volume).

It is not the purpose of this essay to put the present incumbents into any perspective. However, it would be remiss to ignore group contributions, particularly of the CSIRO. These began with the classic land system studies (Christian and Stewart, 1953) and the many that followed, with now more specific studies from various divisions. Other small groups include an Australian National University and University of N.S.W. team working on the South Alligator (Woodroffe *et al*, 1986; 1987) and the Office of the Supervising Scientist, together with CSIRO, and the Universities of Wollongong and Sydney, on the East Alligator of the Northern Territory (East, 1986; Pickup *et al*, 1987).

V.   THIS BOOK

This section comments briefly on the content of the following essays. They have been ordered and grouped in what is hoped to be a fairly logical sequence. However there are no formal sections. Two of the essays are on runoff; three are on various approaches to denudation; three more are on evolution and past alluviation; four are concerned with channels, changes and adjustments to both natural and human-induced events; and four are on human impacts.

Chapters 2 and 3 deal mainly with runoff. The first, by Finlayson and McMahon, compares runoff and its variability in Australia with the rest of the world to demonstrate that our rivers are perhaps different. Regional variations are also discussed. Bonell in the next focuses on runoff from the very wet tropical forests and adjacent cleared land in north-east Queensland.

Chapters 4 to 6 deal with three very different approaches to denudation. Rieger and Olive review the problems of traditional black-box determination of denudation through sampling at a gauging site. Loughran, Campbell and Elliott show how the tracer Caesium 137 has been used to locate sources of erosion within different parts of a catchment, as well as its role in working out rates of sedimentation in areas of accumulation. Finally, Pickup discusses the problems of modelling erosion and deposition, and presents a method using adjusted Landsat information to model STF (scour-transport-fill) sequences in the rangelands of central Australia.

Landscape evolution in the Quaternary and earlier together with past alluviation are considered in Chapters 7 to 9. Wyrwoll, in the only essay from Western Australia, is more concerned with Quaternary sediment architecture at three spatial levels, rather than with dates. In contrast, Nanson, Young, Price and Rust have many dates to demonstrate the relatively great antiquity of sediments in the Channel Country of western Queensland. They are also able to show a two-stage contemporary morphology, with mud braids as part of megaflow sequences and anastomosing channels as part of low flows. Pickup, Allan and Baker combine ancient surfaces, proto-gorges and their meanders, the present gorge and its slack-water deposits in their study of the Finke in central Australia. The evolution of the gorges provides a backdrop for the assessment of some of the more contemporary megafloods through this system.

The next four chapters (10 to 13) are perhaps a more tenuous group concerned with channels, their morphologies and adjustments to regime shifts and catastrophic events, and the impact of secular climatic changes on flows. All are set in New South Wales, with two east of the divide in coastal valleys and two to the west in the Murray-Darling. The Nanson-Erskine essay is concerned with the broader assessment of channel and floodplain morphologies in coastal valleys, together with the impacts of both catastrophic and other changes. Erskine and Warner focus more narrowly on alternating flood- and drought-dominated regimes, the evidence for them in runoff records and examples of their morphological impacts in the Hunter and Nepean rivers. The Riley essay is the first attempt to document the last regime shift in the mid 1940s from the annual runoff records for the NSW part of the Murray Darling. This work is at an early stage with the impacts of regulatory structures on cumulative changes from the means yet to be considered. Page, working on an 80 km reach of the Murrumbidgee downstream of Wagga Wagga, demonstrates contemporary relations between bankfull capacity and dis-charge levels, which have been somewhat attenuated by dams. Very slow rates of channel change in this backwater system are very different to those demonstrated in the high-energy coastal streams.

The last four chapters (14 to 17) show something of human impacts in both catchments and channels. Olive and Rieger report on their work in the forests of Eden, where they have combined with the NSW Forestry Commission for a long time in helping to understand impacts of different logging practices. East and colleagues attempt to see what lessons can be learned from a former uranium mine in the hot, wet and dry tropics of the North, where rehabilitation was minimal. This will help to refine strategies for the eventual closure of the large Ranger mine to minimise the long-term impacts on the fragile wetlands of the Kakadu National Park. Urban impacts on runoff and erosion were studied in the city of Armidale on the New England Tablelands by Neller. The final essay by Warner and Bird is essentially a NSW-Victoria view of direct human impacts on channels. This includes the effects of dams, weirs, channel improvements, flood-mitigation structures and extractive industries.

REFERENCES

Abrahams, A.D. (1972) Drainage densities and sediment
    yields in eastern Australia. *Aust. Geog. Studies*,
    10, 19-41.
Andrews, E.C. (1903) An outline of the Tertiary history
    of New England. *Rec. Geol. Surv. NSW*, 7, 140-216.
Andrews, E.C. (1910) Geographical unity of Eastern
    Australia in Late- and Post-Tertiary time. *J. Roy.
    Soc. NSW*, 67, 251-350.
Atkinson, G. (1987) A review of soil and geological maps
    of the Nepean river terraces, NSW. *Aust. Geog.*, 18,
    124-136.

Bishop, P. (1982) Stability or change: a review of ideas on ancient drainage in eastern NSW. *Aust. Geog.*, 15, 219-230.

Bowler, J.M. (1986) Quaternary landform evolution. In D.N. Jeans, (ed). *Australia - a Geography*, Vol. 1, 117-147.

Butler, B.E. (1958) Depositional systems of the Riverine Plain of south-eastern Australia in relation to soils. *CSIRO Soil Publ.*, 10.

Butler, B.E. (1961) Ground surfaces and the history of the Riverine Plain. *Aust. J. Sci.*, 24, 39.

Butler, B.E. (1967) Soil periodicity in relation to landform development in SE Australia. In J.N. Jennings and J.A. Mabbutt (eds.) *Landform Studies from Australia and New Guinea*, ANU Press, Canberra, 231-255.

Browne, W.R. (1975) Frank Alfred Craft: 1906-1973. *Aust. Geog.*, 13, 1-3.

Browne, W.R. and Raggatt, H.G. (1935) Notes on buried rivers in eastern Australia. *Aust. Geog.*, 2, 24-30.

Christian, C.S. and Stewart, G.A. (1953) General report on survey of Katherine-Darwin region, 1946. *CSIRO Land Res. Ser.*, No. 1.

Cornish, P.M. (1977) Changes in seasonal and annual rainfall in New South Wales. *Search*, 8, 38-40.

David, T.W.E. (1877) Geology of the Vegetable Creek tin-mining field. *Mem. Geol. Surv. NSW*, No. 1.

Davies, J.L. and Williams, M.A.J. (eds.) (1978) *Landform Evolution in Australasia*, ANU Press, Canberra.

Douglas, I. (1973) Rates of denudation in selected small catchments in eastern Australia. *Univ. Hull Occas. Papers in Geog.*, 21.

Douglas, I. (1977) *Humid Landforms*, ANU Press, Canberra.

Douglas, I. (1983) *The Urban Environment*, Arnold, London.

Dury, G.H. (1964) Principles of underfit streams. *U.S.G.S. Prof. Paper 452-A*.

Dury, G.H. (1970) A re-survey of part of the Hawkesbury River, NSW, after 100 years. *Aust. Geog. Studies*, 8, 121-132.

Dury, G.H. (1976) Discharge prediction, present and former, from channel dimensions. *J. Hydrol.*, 30, 219-245.

Dury, G.H., Hails, J.R. and Robbie, H.B. (1963) Bankfull discharge and the magnitude-frequency series. *Aust. J. Sci.*, 26, 123-124.

East, T.J. (1986) Geomorphological assessment of sites and
    impoundments for the long term containment of uranium
    mill tailings in the Alligator Rivers region. *Aust.
    Geog.*, 17, 16-21.

Gentilli, J. (ed.) (1971) *Climates of Australia and New
    Zealand*, Elsevier, Amsterdam.

Hall, L.D. (1927) The physiographic and climatic factors
    controlling the flooding of the Hawkesbury River at
    Windsor. *Proc. Linnean Soc. NSW*, 52, 133-152.

Hickin, E.J. (1969) A newly identified process of point bar
    formation in natural streams. *Am. J. Sci.*, 267,
    999-1010.

Hickin, E.J. (1970) The terraces of the Lower Colo and
    Hawkesbury drainage basins, NSW. *Aust. Geog.*, 11,
    278-287.

Hills, E.S. (1939) The physiography of North-Western
    Victoria. *Proc. Roy. Soc. Vict.*, 51, 112-139.

Jeans, D.N. (ed.) (1977) *Australia - a Geography*. Sydney
    U.P.

Jeans, D.N. (ed.) (1986) *Australia - a Geography, Vol. 1,
    The Natural Environment*, Sydney U.P.

Jeans, D.N. and Davies, J.L. (1984) Australian geography -
    1972-1982. *Aust. Geog. Studies*, 22, 3-35.

Jennings, J.N. and Mabbutt, J.A. (eds.) (1967) *Landform
    Studies from Australia and New Guinea*, ANU Press,
    Canberra.

Jennings, J.M. and Mabbutt, J.A. (1977) Physiographic
    outlines and regions. In D.N. Jeans (ed.) *Australia -
    a Geography*, Sydney, U.P., 38-52.

Jennings, J.N. and Mabbutt, J.A. (1986) Physiographic
    outlines and regions. In D.N. Jeans (ed.) *Australia -
    a Geography*, Vol. 1, Sydney U.P., 80-96.

Josephson, J.P. (1885) History of floods in the Hawkesbury
    River. *J. Roy. Soc. NSW*, 19, 97-107.

Joyce, E.B. (1987) The publications of E.S. Hills.
    Aust. Geog. Studies, 25, 102-109.

King, L.C. (1962) *Morphology of the Earth*. Oliver and
    Boyd, London.

Langford-Smith, T. (1960) The dead river systems of the
    Murrumbidgee. *Geog. Rev.*, 50, 368-389.

Langford-Smith, T. (ed.) (1980) *Silcrete in Australia*,
    Geog. Dept., Univ. New England.

Maiden, J.H. (1902) The mitigation of floods in the Hunter
    River. *J. Roy. Soc. NSW*, 36, 107-131.

Mabbutt, J.A. (1977) *Desert Landforms*, ANU Press,
    Canberra.

Nanson, G.C. (1986) Episodes of vertical accretion and catastrophic stripping: a model of disequilibrium flood plain development. *Geol. Soc. Amer. Bull.*, 97, 1467-1475.

Nanson, G.C. and Young, R.W. (1987) Comparison of thermoluminescence and radiocarbon age determinations from Late-Pleistocene alluvial deposits near Sydney, Australia. *Quat. Res.*, 27, 263-269.

Nanson, G.C., Young, R.W. and Stockton, E.D. (1987) Chronology and palaeoenvironment of the Cranebrook terrace (nr. Sydney) containing artefacts more than 40,000 years old. *Archaeol. Ocean.*, 22, 72-78.

Ollier, C.D. (1986) Early landform evolution. In D.N. Jeans (ed.) *Australia - a Geography*, Sydney U.P., 97-116.

Pels, S. (1964) The present and ancestral Murray River system. *Aust. Geog. Studies*, 2, 119-119.

Pels, S. (1966) Late Quaternary chronology of the Riverine Plain of Southeastern Australia. *J. Geol. Soc. Aust.*, 13, 27-40.

Pickup, G. (1986) Fluvial landforms. In D.N. Jeans (ed.) *Australia - a Geography*, Sydney UP, 148-179.

Pickup, G., Wasson, R.J., Warner, R.F., Tongway, D. and Clark, R.L. (1987) A feasibility study of geomorphic research for long term management of uranium mill tailings. *CSIRO, Inst. Nat. Res. and Environ. Div. Rept.*, 87/2.

Pillans, B. (1983) Canberra, twenty million years on. *Aust. Geog. Studies*, 21, 92-101.

Riley, S.J. (1977) Some downstream trends in the hydraulic, geometric, and sedimentary characteristics of an inland distributary system. In K.J. Gregory (ed.) *River Channel Changes*, Wiley, London, 337-352.

Riley, S.J. (1981) The relative influence of dams and secular climatic change on downstream flooding, Australia. *Water Res. Bull.*, 17, 361-366.

Schumm, S.A. (1968) River adjustment to altered hydrologic regimes - Murrumbidgee river and palaeochannels, Australia. *U.S.G.S. Prof. Paper 598*.

Schumm, S.A. (1977) *The Fluvial System*, Wiley, New York.

Singh, G., Opdyke, N.D. and Bowler, J.M. (1981) Late Cainozoic stratigraphy, palaeomagnetic chronology and vegetational history of Lake George, NSW. *J. Geol. Soc. Aust.*, 28, 435-452.

Spate, O.H.K. and Jennings, J.N. (1972) Australian geography - 1951-1971. *Aust. Geog. Studies*, 10, 133-140.

Taylor, T.G. (1911)   The physiography of eastern Australia. *Bur. of Met. Bull.*, 8.

Twidale, C.R. (1976)   *Analysis of Landforms*, Wiley, Sydney.

Voisey, A.H. (1942)   The Tertiary land surface in southern New England, NSW. *J. Roy. Soc. NSW*, 76, 82-85.

Walker, P.H. and Coventry, R.J. (1976)   Soil profile development in some alluvial deposits of eastern New South Wales. *Aust. J. Soil Res.*, 14, 305-317.

Warner, R.F. (1972)   River terrace types in the coastal valleys of NSW. *Aust. Geog.*, 12, 1-22.

Warner, R.F. (1986)   Hydrology. In D.N. Jeans (ed.) *Australia - a Geography*, Sydney UP, 49-79.

Warner, R.F. (1987a)   Spatial adjustments to temporal variations in flood regime in some Australian rivers. In K. Richards (ed.) *River Channels, Environment and Process*, Blackwell, Oxford, 14-40.

Warner, R.F. (1987b)   The impacts of alternating flood- and drought-dominated regimes on channel morphology at Penrith, NSW, Australia, *IAHS Publ.*, No. 168, 327-338.

Woodroffe, C.D., Chappell, J.M.A., Thom, B.G. and Wallensky, E. (1986)   Geomorphological dynamics and evolution of the South Alligator tidal river and plains, Northern Territory. *ANU North Aust. Res. Unit, Mangrove Mono.*, No. 3.

Woodroffe, C.D., Thom, B.G., Chappell, J.M.A., Wallensky, E., Grindrod, J. and Head, J. (1987)   Relative sea level in the South Alligator River region, North Australia, during the Holocene. *Search*, 18, 198-200.

Woodyer, K.D. (1968)   Bankfull frequency in rivers. *J. Hydrol.*, 6, 114-142.

Young, R.W. (1970)   The patterns of some meandering valleys in NSW. *Aust. Geog.*, 1, 269-277.

Young, R.W. (1978)   The study of landform evolution in the Sydney region: a review. *Aust. Geog.*, 14, 77-93.

Young, R.W. (1981)   Denudation history of the south-central uplands of NSW. *Aust. Geog.*, 15, 77-88.

2

Australia *v* the World:
A Comparative Analysis of Streamflow Characteristics

B.L. Finlayson

Department of Geography
University of Melbourne
Parkville,   Victoria

T.A. McMahon

Department of Civil and Agricultural Engineering
University of Melbourne
Parkville,   Victoria

I.   INTRODUCTION

Australia's position as the driest continent is undisputed and most discussions of Australian water resources begin with a statement to this effect. Aside from this problem of widespread aridity it has long been assumed that Australian rivers behave in much the same way as rivers throughout the world and that the rules governing their behaviour are therefore transferable. The basic scientific principle that hydrologic processes in Australia follow a universal set of physical laws is not questioned. However, there now appears to be good reason to suggest that the particular outcome in the Australian environment of the interaction between process variables and catchment characteristics is a hydrological system which differs from that of the continental areas of the northern hemisphere where hydrology as a science had its origin. The theory and practice of hydrology in Australia owes much to this overseas experience. Most hydrological education in Australia is based on textbooks from Britain or North America;

there are no Australian textbooks on hydrology though some
general accounts of water resources are available (eg Munro,
1974; Pigram, 1986).

One clear example of the problems inherent in this
assumption that Australia, hydrologically speaking, is just
a somewhat drier version of the United Kingdom or the United
States can be seen in the attempts to apply in Australia
early stochastic models of streamflow developed in the
Harvard Water Program (Thomas and Burden, 1963). These
models were tested for streams with low variability in the
United States and were found to be unsatisfactory when
applied in Australia by several water management authorities.
These experiences caused a severe setback to the acceptance
of stochastic models for streamflow data generation in
Australia and have pointed to the need for a better understanding of global hydrology so that regions can be
established for model transferability. Indeed it has recently
been suggested (Prof W. Williams, pers. comm.) that when
the globe as a whole is considered, Australia contains a
more representative range of hydrologic environments than
those northern hemisphere heartlands of scientific hydrology.

Published reviews of global streamflow characteristics
are relatively few in number and based either on data with
limited geographic range or on average conditions derived
from water balance calculations. The water balance studies
confirm Australia's position as the driest continent despite
the fact that mean runoff depth for the southern hemisphere
(316 mm) exceeds that of the northern hemisphere (243 mm)
(Baumgartner and Reichel, 1975). Alyushinskaya *et al* (1977)
also described runoff variability for each of the continents.
Without specifying the origin of the data, they reported
that Australia had the highest coefficient of variation of
total annual runoff of 0.43 compared to a world mean of
only 0.03.

Kalinin (1971) analysed runoff data for 137 stations
worldwide. Unfortunately his data set included only eight
rivers from the southern hemisphere, four from Australia
and four from South America so that while he was able to
analyse his data in a spatial context for the northern
hemisphere, he was unable to extend this to the southern
hemisphere. The first analyses of a southern hemisphere
data set comparable to the work of Kalinin in the northern
hemisphere were carried out by McMahon (1973, 1975a, 1975b,
1977, 1978, 1979). McMahon's work indicated that
Australian streamflow did not necessarily follow the general
relationships established for the northern hemisphere by
Kalinin and that the points of difference mainly concerned

measures of variability. This raised the possibility that Australian rivers were different to those of other continents though one possible interpretation of the data was that the differences were between the northern and southern hemispheres. Brown (1983) states the general scientific position on this matter when he says: "There would not appear to be any *a priori* reason why the runoff from Australian catchments should be more variable than that from catchments of comparable sizes in other parts of the world with similar rainfall and catchment characteristics" (p.35).

In this essay intercontinental comparisons of streamflow statistics are made using a global data set. This is followed by a summary of the regional variations of Australian streamflow characteristics. Finally the implications of these characteristics for Australian hydrology and river management are considered.

II. THE DATA

The distribution of gauging stations used in this comparative study is shown in Figure 1 and the numbers for each continental area are shown in Tables I and III. The records consist of monthly and annual flows for the whole of the available period of record at each station and a separate data set consisting of the peak instantaneous discharge in each year. For the monthly and annual flow data there are 156 Australian stations out of a total of 974 worldwide with a combined record length of 32,100 station years. Peak flow records are available for 169 Australian stations out of a total of 934 worldwide with a combined record length of 28,000 station years. Rainfall data used in these analyses were obtained on magnetic tape from the National Centre for Atmospheric Research (NCAR, Boulder, Colorado).

All the data have been processed in water years. A description of the compilation of the data set and its structure can be found in Finlayson *et al* (1986). Measured and derived values used in this discussion and their abbreviations are as follows:

MAR Mean annual runoff (mm)
C_{vr} Coefficient of variation of annual flows
    calculated as the standard deviation divided by
    the mean

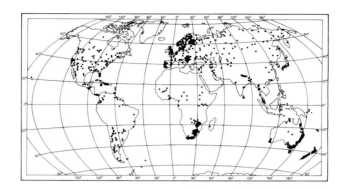

FIGURE 1. Distribution of gauging stations in the world data set.

C_{vp}	Coefficient of variation of annual precipitation
C_{vpe}	Coefficient of variation of annual effective precipitation
τ_{80}	Storage capacity divided by mean annual flow required to meet 80% of draft at 95% reliability
$\bar{Q}$	Specific mean annual flood ($m^3s^{-1}km^{-2}$)
$\bar{q}$	Mean annual flood (m^3s^{-1})
I_v	Standard deviation of annual peak instantaneous flows in the log domain
$Q_{100}/\bar{Q}$	The 100 year recurrence interval flood divided by the mean annual flood.

The distribution of data as shown in Figure 1 is somewhat uneven and there are a number of reasons for this. Major desert areas (such as the Western Plateau of Australia) are virtually devoid of streams and/or stream gauging stations. It has not been possible to obtain data from a number of countries. Streamflow records have only been used when certain minimum requirements have been met in terms of record length and freedom from flow regulation and this has also limited coverage. Despite these problems, the present data set is the most comprehensive global coverage available to date.

III. INTERCONTINENTAL COMPARISONS

A. Annual Runoff

   Table I shows annual flow characteristics of the major
continental areas stratified by catchment size. For each
catchment size class, the number of streams, average record
length and average catchment size is also shown. Australia's
position as the driest continent is confirmed by the MAR
values, especially for the larger catchments. The
Australian data for smaller catchments (<1000 km^2) are
derived mainly from the more humid southeastern corner of
the continent as shown in Figure 1. The other two values
listed in Table I are both measures of variability and here
some important differences between the continents appear.
   Australia (AUS) and Southern Africa (SAF) both exhibit
variabilities that are in excess of the world average across
the range of catchment sizes. Since it is well known that
rainfall variability increases as annual total decreases
(Conrad, 1941) the behaviour of these two continental areas
may simply reflect their aridity. At the annual level MAR
can be used as a surrogate climatic indicator and C_{vr} is
plotted against MAR in Figure 2. Two pertinent facts
emerge from this figure. The AUS and SAF data tend to plot
above the rest of the world (ROW) across the whole range
of MAR's while following the same trend of decreasing C_{vr}
with increasing MAR. This indicates that the differences
shown in the average values in Table I are real and not an
artifact of the averaging.
   The climatic relations of variability can also be
investigated by stratifying the data using Köppen-Geiger
climate zones as shown in Table II. Since large catchment
areas are likely to stretch across more than one climate
zone only catchments <10,000 km^2 are considered here. In
order to maximise sample sizes in each climate zone, AUS
and SAF catchments are combined (ASAF) and compared with
ROW. In nearly all cases where sample sizes are sufficiently
large for reliable comparison the C_{vr}'s for ASAF are greater
than ROW by a factor of the order of 2.0. The exception
is the climate zone Csa where the ASAF value is only 0.7
of the ROW. Only Australian catchments are involved since
there are no Csa climates in Southern Africa. These data
indicate that even where like climates are being compared,
ASAF catchments are still, for the most part, more variable
than those in the rest of the world. The possibility remains,
of course, that this analysis is just revealing a weakness

TABLE I.   Hydrologic Characteristics Based on Annual Streamflow Volumes

		No.	Av. Yrs	Area	MAR	C_{VR}	τ_{80}
World	1	434	28	321	818	0.45	0.79
	2	273	35	3376	542	0.48	0.99
	3	180	36	37674	407	0.37	0.54
	4	87	47	527311	230	0.33	0.46
	5	974	33	55153	612	0.43	0.77
N. Africa	1	0					
	2	3	11	3835	166	0.54	0.86
	3	5	34	48168	190	0.37	0.39
	4	15	30	301913	207	0.25	0.23
	5	23	29	207871	198	0.31	0.35
S. Africa	1	55	26	298	284	0.81	2.38
	2	30	31	3315	102	0.78	1.82
	3	11	30	31235	176	0.70	1.61
	4	4	38	299253	71	0.54	0.98
	5	100	28	16564	209	0.78	2.07
Asia	1	42	16	241	904	0.47	0.81
	2	31	21	4636	793	0.45	1.06
	3	45	14	32163	398	0.30	0.31
	4	25	36	734461	305	0.28	0.29
	5	143	20	139599	616	0.38	0.62
N. America	1	83	23	346	1690	0.31	0.30
	2	54	47	3141	722	0.39	0.54
	3	33	45	36490	507	0.38	0.63
	4	19	61	403495	150	0.35	0.42
	5	189	37	47984	1052	0.35	0.44
S. America	1	11	34	516	637	0.39	0.47
	2	17	34	3537	733	0.33	0.34
	3	21	38	39614	669	0.34	0.42
	4	4	47	1983645	453	0.41	0.96
	5	53	36	166647	666	0.35	0.44
Europe	1	103	34	334	518	0.30	0.28
	2	79	33	3234	515	0.27	0.24
	3	61	47	42500	352	0.31	0.35
	4	17	65	342757	250	0.25	0.19
	5	260	39	33497	460	0.29	0.28
Sth. Pacific	1	40	20	298	1171	0.26	0.24
	2	10	18	2432	1753	0.22	0.15
	3	0					
	4	0					
	5	50	20	725	1287	0.25	0.22
Australia	1	100	35	323	547	0.59	1.10
	2	49	42	3212	208	0.88	2.53
	3	4	46	30275	81	0.98	3.09
	4	3	30	210333	37	1.12	3.50
	5	156	37	4306	416	0.70	1.65

Area Ranges (km²):	1	0-1000	2	1000-10,000
	3	10,000-100,000	4	100,000+
	5	All		

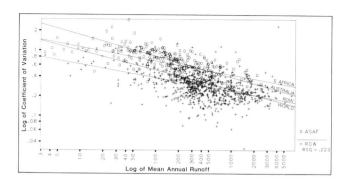

FIGURE 2. The relationship between C_{vr} and MAR for the world data set.

in the climatic classification system which is derived from long term averages and contains no variability component. However, these conclusions are reinforced by the fact that when the comparison is carried out using river regime types the same pattern emerges.

The anomalous character of the Csa climates is of interest in the light of a map published by Leeper (1970) where he applied Conrad's (1941) global relationship between mean annual rainfall and variability to Australia. This showed that most of Australia had higher than expected variability except that part of the southwest of Western Australia and the southeast of South Australia having Csa climates.

On logical grounds it would be expected that variability of annual flows should decrease as catchment area increases. This relationship is shown in Figure 3 for AUS, SAF, ROW and WOR. The correlations, while being statistically significant, are very weak and the general trend is as expected. Australia is a notable exception in this regard. The C_v's are higher than the rest of the world across the whole range of catchment sizes and there is a marked tendency for variability to increase as catchment size increases. This is in contrast to the expected pattern and to the observed pattern for the rest of the world. Southern Africa, which like Australia has higher C_v's across the range of catchment sizes, does not show the same tendency for C_v to increase with catchment size. In this regard Southern Africa is similar to North America, Europe and South America.

The observed relationship between catchment area and C_v for Australia can be explained in terms of the

TABLE II.   Annual Flow Variability Stratified by Catchment Area and Climatic Type

| Climatic region | Australia + Southern Africa | | | Rest of World | | | Ratio of C_v s |
	No. of streams	Average catchment area (km^2)	C_v	No. of streams	Average catchment area (km^2)	C_v	
			Catchments 0-1,000 km^2				
Am	3	370	.42	4	95	.37	1.1
Aw	6	140	.47	55	380	.31	1.5
Bsk	5	350	.93	3	300	.44	2.1
Cfa	14	3340	.79	18	500	.24	3.3
Cfb	64	310	.56	103	290	.27	2.1
Csa	7	350	.50	19	100	.72	0.7
Cwa	16	290	.77	11	310	.33	2.3
Cwb	18	330	.57	6	470	.27	2.1
			Catchments 1,000-10,000 km^2				
Aw	1	8,600	.76	7	2,300	.26	(2.9)
Bsk	9	3,100	.93	9	5,400	.36	2.6
Cfa	26	3,200	.92	40	3,700	.34	2.7
Cfb	18	3,800	.87	32	3,000	.25	3.5
Csb	2	5,700	.61	1	3,200	.50	(1.2)
Cwa	9	2,100	.96	5	5,100	.23	4.2
Cwb	12	3,200	.64	7	4,700	.41	1.6

(-)   Parenthesis indicates sample size very small
0.7   Underline indicates ratio less than unity

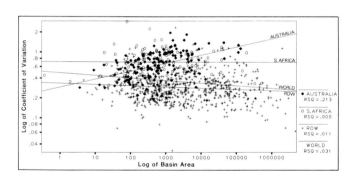

FIGURE 3.   The relationship between C_{vr} and catchment area for the world data set.

distribution of climatic zones on the Australian continent. Humid climatic zones parallel the coast in a relatively thin strip around the northern, eastern, south-eastern and south-western coasts and it has already been shown that variability increases as the climate becomes more arid (Fig. 2).   Large catchments in Australia must include part of the more arid interior and therefore have relatively high C_v's. Typical examples of their C_v's are: Burdekin 0.98, Fitzroy 1.12, Diamantina 1.12 and the Ord 0.79.

B. Annual Floods

The characteristics of the annual peak instantaneous discharge data are summarised in Table III and here also when variability (I_v) or extreme behaviour ($Q_{100}/\bar{Q}$) are considered Australia is distinctly different to the world average and to all the other continents except Southern Africa. An interesting result from these data is the relationship between $\bar{q}$ and catchment area (Fig. 4). The correlation is strong (AUS has the lowest R^2 of 47%) and the nature of the relationship is remarkably similar for all continents.

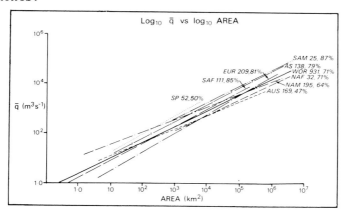

FIGURE 4. The relationship between $\bar{q}$ and catchment area for all stations (WOR), and for each continent individually. The lines are least squares regression lines and in each case the number of stations and the coefficient of determination are shown. Abbreviations for continents are: South America, SAM; Asia, AS; Europe, EU; South Pacific, SP; Northern Africa, NAF; and North America, NAM.

The features observed for annual runoff in the relationship between variability and catchment area are repeated when annual floods are considered. As can be seen from Table III both AUS and SAF have higher I_vs across the range of catchment sizes than the other continents and for AUS, I_v increases as catchment area increases. The ratio $Q_{100}/\bar{Q}$ is also a measure of the variability of flood behaviour and

TABLE III.   Hydrologic Characteristics Based on Annual
Peak Discharges

		No.	Av. Yrs	Area	$\overline{Q}$	I_V	$Q_{100}/\overline{Q}$
World	1	423	27	301	0.80	0.35	3.72
	2	244	32	3534	0.23	0.31	3.90
	3	176	31	36270	0.08	0.26	3.62
	4	88	42	548994	0.03	0.19	2.86
	5	931	30	59812	0.44	0.31	3.67
N. Africa	1	0					
	2	7	15	3721	0.12	0.34	4.07
	3	10	30	34065	0.04	0.23	2.75
	4	15	35	566690	0.02	0.08	1.45
	5	32	29	277095	0.05	0.18	2.43
S. Africa	1	72	26	209	0.46	0.60	5.01
	2	25	31	3482	0.15	0.47	6.43
	3	10	29	28249	0.08	0.39	12.43
	4	4	27	152020	0.01	0.40	5.66
	5	111	27	8943	0.34	0.54	6.02
Asia	1	40	16	262	0.58	0.35	3.75
	2	31	19	4110	0.40	0.23	3.11
	3	43	14	31832	0.11	0.22	2.74
	4	24	42	691433	0.04	0.16	3.41
	5	138	21	131167	0.30	0.25	3.23
N. America	1	111	24	266	1.36	0.26	3.47
	2	37	34	4182	0.30	0.27	4.44
	3	28	32	32505	0.08	0.27	3.97
	4	19	39	482157	0.02	0.16	2.17
	5	195	28	52592	0.85	0.25	3.60
S. America	1	2	17	808	0.53	0.18	5.63
	2	6	25	3771	0.23	0.16	2.07
	3	13	29	46767	0.11	0.14	2.02
	4	4	48	1557192	0.04	0.10	1.57
	5	25	30	274439	0.16	0.14	2.25
Europe	1	61	36	438	0.21	0.16	2.17
	2	76	36	3302	0.12	0.17	2.21
	3	56	41	42857	0.06	0.18	2.27
	4	16	57	363555	0.03	0.14	1.97
	5	209	39	40644	0.12	0.17	2.20
Sth. Pacific	1	40	23	295	1.32	0.24	2.92
	2	12	22	2284	0.86	0.16	2.49
	3	0					
	4	0					
	5	52	23	754	1.21	0.22	2.82
Australia	1	97	32	332	0.68	0.44	4.29
	2	50	40	3322	0.16	0.55	5.84
	3	16	42	29597	0.05	0.66	6.49
	4	6	41	233668	0.01	0.69	7.78
	5	169	36	12272	0.45	0.50	5.08

Area Ranges (km²):   1   0-1000   2   1000-10,000
                     3   10,000-100,000   4   100,000+
                     5   All

in this case Q_{100} has been calculated assuming that the annual series follows a power normal distribution (Chandler *et al*, 1978). It is listed in Table III and plotted against catchment area in Figure 5. Although the correlations are weak, they are statistically significant and the distribution of data points (Fig. 5) confirms the reality of the difference between ASAF and the other continents.

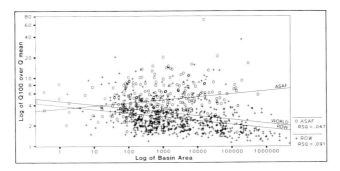

FIGURE 5. The relationship between $Q_{100}/\bar{Q}$ and catchment area for the world data set.

The variability of the annual flood series, I_v, is stratified by climatic zones in Table IV. The pattern revealed in Table II for C_v is repeated in Table IV. Australia and Southern Africa are, on average, about twice as variable as comparable climatic zones in the rest of the world. In this case the exceptions to the rule are the Csa and catchments <1000 km^2 in BSk. The explanation offered earlier in relation to C_v can be applied also in this case and it should be noted that BSk climates lie on the arid margin of the Csa zones in Australia.

IV. REGIONAL CHARACTERISTICS OF AUSTRALIAN STREAMS

A. Annual Runoff

As can be seen from the distribution of Australian gauging stations in Figure 1 the data coverage is spatially uneven. Not all possible gauging stations have been included in this data set but, given the requirements of length of record and lack of regulation, any attempt to increase the number of stations would only result in more data in those areas already well covered. There are also large areas of Australia which lack integrated river systems

TABLE IV.   Peak Discharge Variability Stratified by Catchment Area and Climatic Type

Climatic region	Australia + Southern Africa			Rest of World			Ratio of I_v s
	No. of streams	Average catchment area (km²)	I_v	No. of streams	Average catchment area (km²)	I_v	
Catchments 0-1,000 km²							
Am	7	360	.31	5	92	.13	2.4
Bsh	12	250	.80	1	660	.21	(3.8)
Bsk	7	250	.47	7	91	.62	0.8
Cfa	17	390	.52	27	280	.21	2.5
Cfb	49	340	.39	62	350	.18	2.2
Csa	12	280	.29	20	110	.46	0.6
Csb	15	160	.29	3	150	.21	1.4
Cwa	30	180	.51	8	310	.26	2.0
Cwb	18	290	.32	4	480	.23	1.4
Catchments 1,000-10,000 km²							
Am	1	1,900	.36	5	3,400	.17	(2.1)
Bsk	8	3,400	.75	4	4,400	.40	1.9
Cfa	27	3,500	.48	18	3,500	.24	2.0
Cfb	15	3,800	.50	29	2,900	.19	2.6
Csa	2	6,800	.41	6	4,500	.46	(0.9)
Csb	5	4,200	.36	1	3,200	.26	(1.4)
Cwa	7	2,200	.53	5	4,600	.20	2.7
Cwb	10	3,600	.38	6	4,700	.17	2.2

(-)   Parenthesis indicates sample size very small
0.8   Undreline indicates ratio less than unity

either because of climatic aridity, as is the case over much of the Western Plateau Drainage Division, or geology, as is the case for the Nullarbor Plain.

The data problem in Australian runoff has recently been highlighted by Brown (1983) who, in a review of Australia's surface water resources, produced updated estimates of mean annual runoff by drainage divisions which he compared with similar estimates made by earlier workers. Brown's estimates and those of the AWRC (1976) are shown in Table V. The differences are attributed largely to an increase in the number of gauged catchments for which records were available in 1982, though changes in the procedures for estimating runoff from ungauged catchments were also a factor. Given the high temporal variability of runoff in Australia as discussed above, estimates of mean annual runoff based on short periods of record are likely to be unstable. The distribution of runoff depth in Australia is shown in Figure 6 on which the drainage divisions are also shown for ease of comparison with the data of Table V.

Some of the highest annual runoff variabilities in the world are found in Australian streams (see Figs. 2 and 3). The distribution of runoff variability has been mapped by the AWRC (1978) and McMahon (1978). The AWRC have chosen

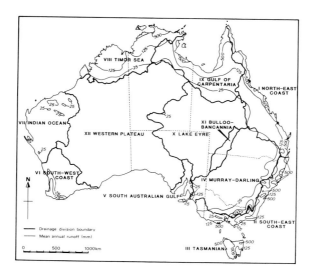

FIGURE 6. Mean annual runoff in Australia and the national drainage divisions. (After Brown, 1983).

TABLE V. Estimates of the Mean Annual Runoff from Australia by Drainage Divisions as shown on Figure 6.

	Estimate of mean annual runoff ($10^6 m^3$) made in the year indicated						
		1975			1982		Change from 1975 %
	gauged	ungauged	total	gauged	ungauged	total	
I	32,200	50,300	82,500	47,200	44,300	91,500	10.9
II	30,800	8,600	39,400	32,700	12,700	45,400	15.2
III	31,400	18,400	49,800	33,500	19,900	53,400	7.2
IV	21,900	300	22,200	22,200	500	22,700	1.8
V	360	620	980	400	600	1,000	0.0(c)
VI	4,800	2,500	7,300	4,500	2,200	6,700	-9.6
VII	2,600	1,500	4,100	3,400	600	4,000	-4.9
VIII	27,900	46,400	74,300	31,500	49,700	81,200	9.2
IX	3,300	54,900	58,200	51,800	78,700	130,500	124.0
X	1,600	1,700	3,300	1,600	1,700	3,300	0.0(c)
XI	460	80	540	500	100	600	+7.4
XII							
	157,400	185,200	342,620	229,300	211,000	440,300	+28.6

to map a variability index calculated as the difference between the 10th percentile and the 90th percentile divided by the median. They argue that this is the most appropriate measure of variability to use because of the skewness of the Australian runoff data. McMahon (1982) has shown that Australian streams have a mean skewness value of 1.29 (range -0.50 to 3.76) compared with 0.47 (-0.33 to 2.02) for the world data set used in that study. In the same study he

reported that while some 70% of streams in the world data
set had normally distributed annual flows, for Australia this
figure was only 30%.

Despite this, the most useful measure of variability
is the coefficient, C_v, defined above.   Figure 7 is based on
McMahon's (1978) map of the distribution of C_v in Australia.
The dominant feature of the pattern (which is very similar
to that of the AWRC map) is the relationship between streamflow variability and climatic aridity.   Pockets of
particularly high C_v (even for Australia) occur in the
central highlands of Queensland and on the western slopes
of central NSW (Fig. 7).   These have never been studied in
detail but McMahon (1978) suggests that they may be related
to regional orographic effects.

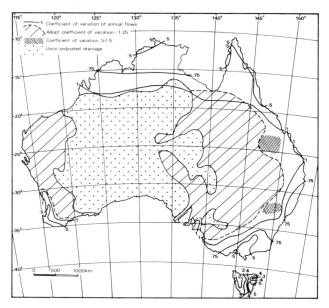

FIGURE 7.   The distribution of C_{vr} in Australia.

B.   *Annual Floods*

A general relationship exists between the variability
of floods and the variability of annual flows and it has
already been shown (Figs. 3 and 5) that Australia's anomalous
characteristics are similar for both measures.   While no
systematic attempt has been made to map flood behaviour for
Australian rivers, it would appear that variability of
the annual flood series follows a similar spatial pattern
to the variability of annual flows as shown in Figure 7.

The timing and relative magnitudes of the first and second ranking floods of record for selected Australian rivers have been compiled by McMahon (1985) (Fig. 8). In 1954, 1955, 1956, 1959, 1971 and 1974, major floods have been recorded in a number of rivers. As this list suggests, the 1950's was an unusual decade in Australian streamflow. The ratio of the largest flood on record to the mean annual flood varies widely across Australia. Some extreme values and year of occurrence are: Todd River (NT), 21, 1910; Latrobe river (VIC), 14, 1934; Murrumbidgee River (NSW), 10, 1853; Hunter River (NSW), 10, 1955; Torrens River (SA), 10, 1889.

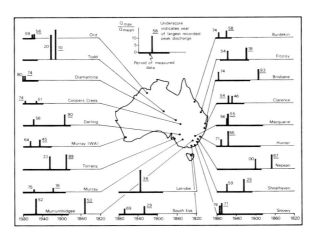

FIGURE 8. Timing and relative ranking of the first and second ranking floods of record for selected Australian rivers (After McMahon, 1985).

Flood magnitude, expressed as the discharge of the 100 year flood in $m^3s^{-1}km^{-2}$, varies widely across Australia (Fig. 9). The drainage divisions yielding the largest floods are North East Coast and the New South Wales part of South East Coast (SEC-E in Fig. 9). Extreme floods in the South West Coast division have peak discharges 1 to 1.5 orders of magnitude less than the South East Coast division. Catchments in the Csa climate zones in Tables II and IV which have C_v's and I_v's below the world average lie in this drainage division. An interesting feature of the data (Fig. 9) is that the envelope curve for the 100 year recurrence interval floods in Australia is almost identical to the envelope curve of the largest observed discharges throughout the world (McMahon, 1985).

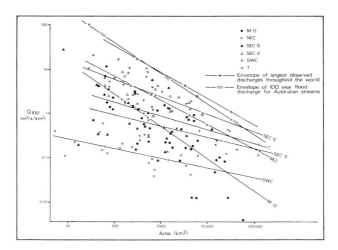

FIGURE 9.   *The relationship between* Q_{100} *and catchment area for the Australian drainage divisions. Least squares regression lines are shown for each division.   Abbreviations are: Murray-Darling, M-D; North-East Coast, NEC; South-East Coast in Victoria, SEC-S; South-East Coast in New South Wales, SEC-E; South-West Coast, SWC; and Tasmanian, T (After McMahon, 1985).*

C. Seasonal Regimes

The global data set discussed earlier has been used to investigate streamflow regimes using cluster analysis. A within-group average clustering method with cosine similarity measure contained in SPSS-X (1986) was used with mean monthly streamflows, arranged so that the first month for each station was the first summer month. Fifteen regime classes were isolated in the global data set and their regime patterns are shown in Figure 10. Ten of the fifteen classes are found in Australia though three (Groups 1, 5 and 6) are poorly represented. Missing from the Australian data are streams with late spring to early summer regimes (Groups 2 - 4) associated with freezing conditions in winter. The mid-autumn class (Group 10) is also not found in Australia and it occurs mainly in those tropical areas which experience an annual double maximum of rainfall. There are no regime classes unique to Australia.

The distribution of river regimes in Australia is shown in Figure 11 where only the seven dominant groups have been mapped. The clustering procedure used is an hierarchical

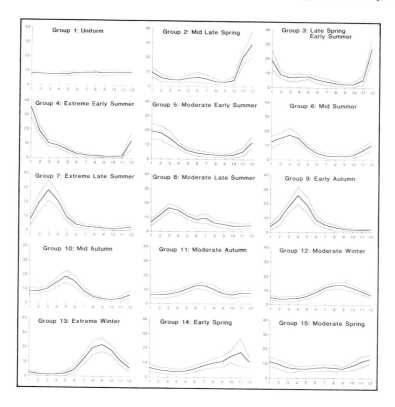

FIGURE 10. *River regime patterns for the global classification derived from the world data set. Heavy lines represent the mean monthly flow for all stations in that group expressed as a percentage of the mean annual flow and the light lines are the one standard deviation limits. 1 is first month of summer on x-axis.*

one and the first order division which appears in the Australian data is between summer and winter dominance. Clearly this is a north-south split. The location of the dividing line on the west coast is indeterminate with the present data, though the map of seasonal distribution of rainfall (Bureau of Meteorology, 1977) puts it in the vicinity of North West Cape. In the east the division occurs farther south on the coast than it does inland with summer dominance persisting as far south as the Hunter Valley on the coast of New South Wales while further inland winter dominance extends as far north as the border between Queensland and New South Wales.

The extreme late summer regime type (Group 7) occurs across the whole of northern Australia in response to the rainfall pattern produced by the southward incursion of the inter-tropical convergence zone and summer tropical cyclone activity. A notable interruption to this pattern occurs

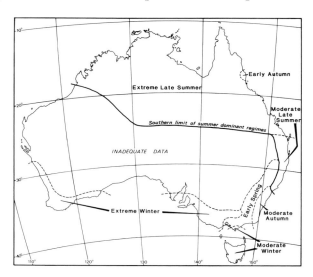

FIGURE 11.   The distribution of seasonal river regimes in Australia.

in north Queensland to the east of the Atherton Tableland. Here orographic rainfall is produced in the winter half year from the trade wind system thus lessening the extreme summer dominance of runoff which characterises the rest of northern Australia.

The extreme winter regime type (Group 13) occurs in the south western corner of Western Australia, the south eastern corner of South Australia, the western district of Victoria and in part of north eastern Victoria. This is the area of winter dominant rainfall related to seasonal shifts in the latitudinal position of the sub-tropical high pressure cells. Tasmania and South Gippsland in Victoria have moderate winter regimes (Group 12).

Three regime types, extreme late summer, extreme winter and moderate winter (Groups 7, 13 and 12 respectively) are areally dominant in Australia and form homogeneous zones (Fig. 11). Four other groups have been mapped on Figure 11 and the remaining three groups found in Australia do not occur in spatially cohesive zones. Early autumn (Group 9) and moderate autumn (Group 11) regimes occur as homogeneous areas in North Queensland and southern coastal New South Wales respectively. The other two zones mapped in Figure 11, moderate late summer (Group 8) and early spring (Group 14), have been defined on the basis of dominant type. In the northern part of the moderate late summer zone there are some streams from Groups 5, 6 and 7 though the zone is occupied exclusively by Group 8 streams south of the New South Wales border. Streams from Group 12 occur scattered throughout the early spring zone. A small group of moderate spring (Group 15) streams occupies the northern end of the early spring zone and the only two uniform type streams in Australia occur in the south of this zone.

V. DISCUSSION

A. *An Explanatory Model*

It would appear, on the basis of the data presented here, that Australia and Southern Africa have more variable annual river flows and annual floods than the rest of the world's continents. The obvious initial explanatory hypothesis is that rainfall in these two continents is more variable than elsewhere. The connection between runoff variability and streamflow variability is difficult to analyse without a time series of catchment rainfalls comparable to the runoff data. In the absence of appropriate rainfall data, the relationship has been investigated using only catchments <10,000 km^2, each one paired with the nearest rain gauge drawn from the NCAR tape. While these data would be inappropriate for water balance studies, rainfall variability is considered to be sufficiently stable areally to allow comparison in this way. This notion is supported by Figure 12 where the relationship between C_{vr} and C_{vp} is shown for various separation distances between rain and stream gauges. Figure 12 also shows that for any given C_{vp} the corresponding C_{vr} is much higher in Australia and Southern Africa than in the rest of the world. The observed differences in streamflow variability cannot be explained by higher rainfall variability in Australia and Southern Africa, but rather it is the case that more of the rainfall variability is transferred to runoff.

McMahon *et al* (1987) have attempted to analyse this relationship further with a single linear storage model using only those rain gauge/stream gauge pairs with station separation of less than 55 km. Australian and Southern

African stations were grouped together and compared with stations from the rest of the world. For each station, one thousand years of annual rainfall data were generated synthetically using a Markov process as modified by the Wilson-Hilferty transformation (Wilson and Hilferty, 1931).

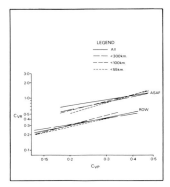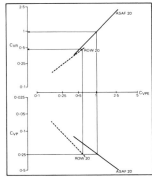

FIGURE 12.   The relationship between C_{vr} and C_{vp} for various separation distances between rain gauge and stream gauge   (After McMahon et al, 1987).

FIGURE 13.   The transfer of variability from rainfall through effective rainfall to runoff for the 20 ASAF stations and the 20 ROW stations used in the model   (After McMahon et al, 1987).

The parameters of the generated data were based on those of the observed data. Annual rainfall totals were converted to annual effective rainfall using a non-linear rainfall-runoff model based on the tanh algorithm used by Boughton (1966). For each catchment the storage effects were mimicked by routing the 1000 years of annual effective rainfall through a conceptual single linear storage model of the form:

$$S = K.R^m \tag{1}$$

where S is catchment storage, R is annual runoff volume and m=1 for a linear model. The storage delay time parameter K was optimised for each catchment so that the time series of annual flows reproduced the observed variability, skewness and autocorrelation.

Initially it was expected that the storage parameter K would differ substantially between the two groups of catchments because of morpho-lithological differences between Australia and Southern Africa and the rest of the world. In fact, as can be seen from Table VI, there are no significant differences between the two groups in terms of K. The important difference is in the transfer of variability from rainfall to effective rainfall. This is summarised graphically in Figure 13. For any given rainfall variability, the variability of effective rainfall is higher for ASAF than ROW and the difference increases as rainfall variability increases. The two groups behave similarly in the transfer of variability from effective rainfall to runoff.

TABLE VI. Median Values of Input Data and Results Parameters for Pairs of Rainfall and Runoff Stations Used in the Storage Model

		ASAF (20)	*ROW* (20)
MAP (mm)		820	800
C_{vp}		0.24	0.16
Runoff coefficient $= \dfrac{MAR}{MAP}$		0.15	0.36
Rainfall excess = MAR (mm)		120	270
C_{vpe}		0.90	0.39
C_{vr}		0.74	0.35
K (yrs)	*(mean)*	0.2	0.3
	(median)	0.2	0.2

This model analysis, though tentative and based on limited data, indicates that the explanation for the anomalous variability of streamflow in Australia (and Southern Africa) lies not in the fact that they are ancient land surfaces of low relief with old soil/landscape systems which were not rejuvenated by continental glaciation during the Pleistocene but in the topographic influences on climate and both Australia and Southern Africa lack the high mountain chains that occur in other continents. The modelled increase in variability between rainfall and effective rainfall is a function of the higher evaporative power of the atmosphere in Australia. A more detailed explanation awaits further research.

B.   *Implications of High Runoff Variability*

The fact that Australian rivers are different has important implications. Channel modifications based on British experience often had disastrous consequences, as when swamp drains became large eroding channels (Warner and Bird, this volume). Government involvement in river management was soon necessary but even then the extreme behaviour of Australian rivers was ignored. To be successful management principles need to take this into account.

The same is true for aquatic ecology. Floral and faunal relationships with hydrology are largely unknown. Little local research has been attempted and extrapolation based on overseas experience does not always work. Another consequence of high runoff variability is its effects on storage (Table I). This, together with the effects of high evaporative losses, means that large storage volumes are necessary for any given level of regulation. The extreme flood behaviour also necessitates larger spillways for dams.

ACKNOWLEDGEMENTS

The research described in this essay was made possible by grants from the Australian Research Grants Scheme.

REFERENCES

Alyushinskaya, N.M., Voskresenskiy, K.P., Grigorkina, T.Ye, Kovzel, A.G., Markova, O.L., Rybkina, A.Ye and Sokolov, A.A. (1977) Global runoff. *Soviet Hydrology*, 16, 127-131.
Australian Water Resources Council (1976) *Review of Australia's Water Resources 1975.*   Dept. of Nat. Res., Canberra.
Australian Water Resources Council (1978) *Variability of Runoff in Australia*. Dept. of Nat. Development, Hydrol. Series No. 11, Canberra.
Baumgartner, A. and Reichel, E. (1975) *The World Water Balance. Mean Annual Global, Continental and Marine Precipitation, Evaporation and Run-off*. Elsevier, Amsterdam.
Boughton, W.C. (1966) A mathematical model for relating runoff to rainfall with daily data. *Civil Eng. Trans. Inst. Engrs. Aust.*, CE8, 83-97.

Brown, J.A.H. (1983) *Australia's Surface Water Resources*. Water 2000, Consultant's Report No. 1, Department of Resources and Energy, Canberra.

Bureau of Meteorology (1977). *Climatic Atlas of Australia*. Department of Science, Canberra.

Chandler, S., Spoila, S.K. and Kumar, A. (1978) Flood frequency analysis by power transformation. *J. Hydr. Div. ASCE*, 104, 1495-1504.

Conrad, V. (1941) The variability of precipitation. *Mon. Weath. Rev.*, 69, 5-11.

Finlayson, B.L., McMahon, T.A., Srikanthan, R. and Haines, A. (1986) World hydrology: a new data base for comparative analyses. *Aust. Inst. Eng.*, National Conference Publication 86/13, 288-296.

Kalinin, G.O. (1971) *Global Hydrology*, Israel Program for Scientific Translations.

Leeper, G.W. (ed.) (1970) *The Australian Environment*, CSIRO and Melbourne University Press, Melbourne, 4th ed.

McMahon, T.A. (1973) Low flow hydrology of selected Australian streams. *Monash Univ. Civ. Eng. Res. Rep.*, 3/73.

McMahon, T.A. (1975a) Predictive use of catchment outputs. In T.G. Chapman and F.X. Dunin (eds.) *Prediction in Catchment Hydrology*, Australian Academy of Science, 371-425.

McMahon, T.A. (1975b) Variability, persistence and yield of Australian streams. *Aust. Inst. Engrs*, National Conference Publication 75/3, 107-111.

McMahon, T.A. (1977) Some statistical characteristics of annual streamflows of northern Australia. *Aust. Inst. Engrs., National Conference Publication*, 77/5, 131-135.

McMahon, T.A. (1978) Australia's surface water resources: potential development based on hydrologic factors. *Civ. Engg. Trans., Aust Inst. Engrs.*, CE20, 155-164.

McMahon, T.A. (1979) Hydrological characteristics of arid zones. IAHS-AISH Publ. No. 128, 105-123.

McMahon, T.A. (1982) *Hydrological Characteristics of Selected Rivers of the World*. UNESCO Inter. Hydrol. Prog. Tech. Documents in Hydrol., Paris.

McMahon, T.A. (1985) Flooding in Australia - yet another review. *Natural Disasters in Australia*. Aust. Acad. Technol. Sci., 9th Invitation symposium, Sydney, Preprint No. 3.

McMahon, T.A., Finlayson, B.L., Haines, A. and Srikanthan, R. (1987)   Runoff variability: a global perspective. In S.I. Solomon, M. Beran, K. Hogg (eds.), IAHS Publ. No. 168, 3-11.

Munro, C.H. (1974)   *Australian Water Resources and Their Development*, Angus and Robertson, Sydney.

Pigram, J.J. (1986)   *Issues in the Management of Australia's Water Resources*, Longman, Sydney.

SPSS-X (1986)   *SPSS-X Information Analysis System*, Release 2.1, SPSS Inc., Chicago.

Thomas, H.A. and Burden, R.P. (1963)   *Op∙rations Research in Water Quality Management*, Harvard Water Resources Group.

Wilson, E.B. and Hilferty, M.M. (1931)   Distribution of Chi-square, *Proc. Nat. Acad. Sci. Wash.*, 17, 684-688.

3

Hydrological Processes and Implications for Land Management in Forests and Agricultural Areas of the Wet Tropical Coast of North-East Queensland

Mike Bonell

Department of Geography
James Cook University
Townsville,   Queensland

I.   INTRODUCTION

The wet tropical coast of north-east Queensland extends between 16°S and 19°S (south of Cooktown to north of Townsville) and incorporated 1,200,000 ha of tropical rainforest prior to European settlement (Cassells and Gilmour, 1978). Nicholson *et al* (1983) estimated that 675,000 ha still remains under tropical rainforest. The loggable area (142,900 ha) amounts to 17 percent of the total rainforest area and is based on a controversial short cutting cycle of 40 to 50 years using selective logging techniques. Apart from settlement, the former rainforest area has been converted to cropping for sugar cane and maize, to horticulture for bananas and to pasture for dairy and beef cattle.

One of the most outstanding features of this wet tropical coast is the marked concentration of high annual rainfall in a few months of the year. Such high rainfalls emanate from a highly active meteorological environment which in turn has repercussions on the storm runoff hydrology and land disturbance by forestry and agricultural activities. This review first emphasises the close links between synoptic climatology, rainfall intensities and storm runoff generation. A principal theme is the way in which tropical rainforest copes with the annual wet season deluge. The

hydrological implications by land disturbance from forestry
and short-rotation agriculture, notably sugar cane, are
then considered. Subsequent attention is concerned with
the options for improved management. A focal point in the
discussion is a set of paired experimental catchments
(17°20'S, 145°58'E) near Babinda.

II.   PHYSICAL ENVIRONMENT

A.   *Location and Physiography*

     The wet tropical coast of north-east Queensland extends
along the mountains from Cairns in the north to Cardwell
in the south. The principal physical features are a series
of ranges, just inland from the coast as part of the Great
Dividing Range which culminate in the peaks of Bartle
Frere (1,622 m) and Bellenden Ker (1,561 m) immediately
to the west of Babinda (Fig. 1).

B.   *Geology and Soils*

     Extensive areas of the wet tropical coast are underlain
by the broad rock types: granites, basalts and metamorphics.
These are supplemented by superficial deposits in the form
of beach ridges and extensive alluvium in the floodplain
areas of the major rivers. A detailed description of the
geology of the area has been supplied by Bureau of Mineral
Resources, eg de Keyser (1964) based on the 1:250,000
Geological Series maps for the area eg Sheets SE55-2 (CAIRNS)
and SE55-6 (INNISFAIL).
     Using the Great Soil Groups of Stace *et al* (1968),
krasnozems are associated with the basalts, whilst the
granites and metamorphics develop both red and yellow
phases, eg red podzolic, yellow podzolic, xanthozem
(Isbell *et al*, 1968). Also red and yellow earths are
commonly identified with colluvial deposits of granitic
origin. The alluvium presents much more varied soil types.
The most comprehensive soils description is given by Murtha
(1986) and incorporates the mid-section of the wet tropical
coast between Innisfail and Tully. A total of 43 soil
series (as defined in Soil Survey Staff, 1975) were
established into seven broad units (beach ridges, basaltic,
metamorphic and granitic origin, well drained soils formed
on alluvium, poorly drained soils on alluvium, organic soils
and peats of the freshwater swamps and soils of the tidal
zone).

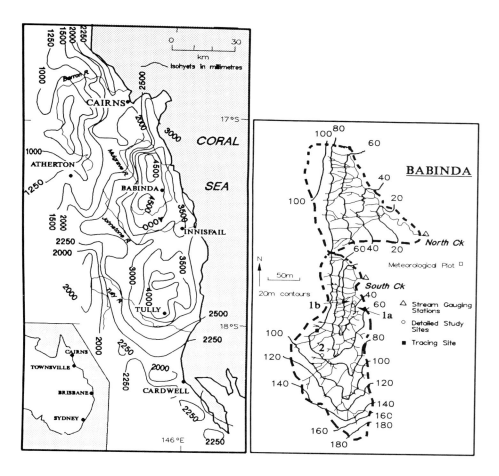

FIGURE 1. *The 30-year mean annual isohyetal map of the wet tropical coast of north-east Queensland and location, experimental sites and physiographic features of the Babinda experimental catchments.*

A preliminary survey of field saturated hydraulic conductivity, K, for krasnozems, red and yellow podzolics and a red earth was undertaken by Bonell *et al* (1983). This was based on 13 tropical rainforest sites incorporating soils of basaltic, metamorphic and granitic origin. Intensive biological activity in the top 0.2 m of all soils makes them capable of accepting most of the prevailing rainfall intensities. Below 0.2 m most soil profiles act as 'impeding layers' to prevailing short term intensities.

For example in the 0.5 to 1.0 m layer, the yellow podzolics and some red podzolics including one of the Babinda catchment runoff plots (site 2) are the least permeable whose log mean K range from 0.5 to 1.25 mm hr^{-1}. An intermediate group of three red podzolic soils ranged from 2.5 to 2.9 mm hr^{-1}. The most permeable were the krasnozems, one of the Babinda catchment runoff plots (site 1a) on red podzolic and red earth derived from colluvium whose log mean K range from 5.4 to 27.5 mm hr^{-1}.

C. Rainfall Climatology

1. General Atmospheric Circulation. The various meteorological systems connected with the general atmospheric circulation of the area were described in detail by Bonell *et al* (1986) and in more abbreviated form elsewhere (Bonell and Gilmour, 1980; Sumner and Bonell, 1986). Thus only a summary of the essential features is given here.

The wet tropical coast is located on the southern fringe of what Ramage (1968) termed the 'maritime continent' which acts as an important global heat engine. The 'maritime continent' has a major influence on the climate of the wet tropical coast by heat energy transfer towards the higher latitudes and, therefore, the atmospheric circulation responsible for rain.

Integrated with this transfer is the interaction between the cross equatorial flow of the equatorial westerlies and the southern hemisphere trade wind easterlies during the summer (December–March) along the thermal low pressure trough. This is known as the *monsoon trough* (Sadler and Harris, 1970) or the *southern monsoon shearline* (McAlpine *et al*, 1983). Cloudiness along this system is highly variable. Some sections are relatively cloud free whilst others have deep convection spread across the wind shear line as a cluster corresponding with a vortex. The maximum cloud zone (MCZ) occurs on the equatorial side of the trough in the equatorial westerlies (Davidson *et al*, 1983).

The wet tropical coast is only occasionally within the MCZ when the monsoon trough advances southward from its normal summer position near 13-14°S. Consequently deep, scattered thunderstorm cells are not the frequent source of rainfall here in contrast with equatorial rainforest areas. More significant is the development and movement of vortices on the monsoon trough in the vicinity of the wet tropical coast. Such systems more commonly occur as tropical depressions (as defined by Lourensz, 1981) but occasionally further development upgrades them as tropical cyclones (as defined by Bureau of Meteorology, 1978). These systems are prodigious rainfall producers and their existence explains why the pattern of deluge on a few days interspaced by minimal activity in the intervening periods occurs.

Another important source of energy transfer from the 'maritime continent' is by the circumpolar upper westerlies which form part of the outflow from the upper anticyclones over the southern monsoon shearline. Low pressure troughs embedded within this upper westerly circulation can occur at any time of the year and are a particularly important source of annual rainfall when summer monsoon activity is weak. Such systems certainly account for the higher rainfalls between April and November.

A third source of rain emanates from the surface easterlies, as part of the Hadley cell return flow to the 'maritime continent'. These winds are associated with a surface ridge of high pressure along the coast, but commonly are too shallow to produce large amounts of rain. For any significant shower activity to occur, the inversion layer has to rise temporally to allow the lower, moist easterlies to gain more depth, which is further enhanced by orographic uplift over the adjacent mountain range.

2. Rainfall

a. Annual, seasonal and daily characteristics with particular reference to the Babinda catchments. Orographic uplift of the prevailing east to southeasterly winds lead to annual rainfall totals well in excess of 2500 mm along large sections of the coastal hinterland (Fig. 1). A distinctive feature is the localised spatial distribution of high annual rainfall with sharp isohyetal gradients to the configuration of the topography in conjunction with the almost perpendicular approach of the prevailing wind onto this section of coastline.

The most outstanding feature of the wet tropical coast is the marked concentration of the high annual rainfalls in a few months of the year due to the circumstances of the atmospheric circulation. For example, of the 4009 mm mean annual rainfall, 1970-83 inclusive, recorded in the Babinda experimental catchments (median 3630 mm, range 2654 mm (1980) to 5496 mm (1973)), 63.5 percent occurred between December and March. Also, the nature of the synoptic meteorology by way of well organised circular disturbances on the monsoon trough means that rain occurring on only a few days makes up a large proportion of the annual rain-

fall. For example, in 1973, 41 percent of the total
precipitation occurred on only 13 days in the Babinda catchments. Daily totals can exceed 250 mm and on several days
a year exceed 100 mm. The highest daily total recorded
in the period 1970-84, was 464 mm in 1976. The average
maximum daily rainfall for raindays exceeding 100 mm was
270.5 mm from 7.1 days per year over the same period (see
Table I, Bonell *et al*, 1986). Furthermore a significant
percentage of the yearly total can also occur on consecutive
days because tropical depressions can be almost stationary
for several days embedded in an active monsoon trough. An
example referring to the summer of 1977 was described by
Bonell and Gilmour (1980). The most outstanding was 2602
mm from 14 days of continuous rain in 1981 (3-17 January)
which amounted to 48.9 percent of the annual total of
5324.5 mm and included three days with rainfalls exceeding
250 mm (Bonell *et al*, 1986).

In common with other similar climatic regions, the
monsoon influence is highly variable in north-east
Queensland. In some years the southward movement of the
monsoon trough is checked for most of the summer by a strong
upper high pressure system over central Australia, which
feeds dry southerly air in the mid-troposphere across the
wet tropical coast. Below-average rainfall is thus recorded
but this pattern can still temporarily break down and allow
monsoonal disturbances to affect the area. In the dry
summer of 1983 the Babinda catchments' January (101 mm)
and February (127 mm) totals were only 13.3 and 18.1
percent of the respective averages and were the lowest on
record (1970-83). A tropical depression however in early
March generated 784.25 mm (4-14 March 1983), including
consecutive daily totals of 176.25 mm (8th) and 239.00 mm
(9th). The significance here is that rainfall still occurs
as short intense bursts over time and results in the same
hydrological management problems that occur in the rainier
wet seasons (Bonell *et al*, 1986).

Daily totals in excess of 100 mm can still occur between
April and mid-June (post-monsoon season, Bonell and Gilmour,
1980) from meridional troughs in the upper westerlies when
21.5 percent of the annual rainfall occurs in the Babinda
catchments. The remaining 15 percent occurs in mid-June
to November when SE stream showers are the dominant rainfall
systems but daily rain amounts from this influence are
small (< 25 mm). Heavier rainfalls however, can still
intermittently be recorded from the passage of upper
westerly troughs.

  b.   Short term rainfall intensities (Babinda catch-

TABLE I. *Mean Annual Runoff/Precipitation for Selected Streams on the Wet Tropical Coast*

Basin/Location of Gauging Station	Area (km^2)	Mean Annual Precipitation (MAP) (mm)	Mean Annual Runoff (MAR) (mm)	% MAR MAP
Babinda Creek				
(Babinda - 17°21'S 145°55'E)	91	4745 (1919-69)	4097 (1926-80)	86.34
Behana Creek				
(Aloomba - 17°8'S 145°50'E)	85	3573 (1919-69)	2081 (1923-71)	58.24
Fisher Creek				
(Nerada - 17°33'S 145°54'E)	16	4400 (1919-69)	2783 (1927-80)	63.25
North Johnstone River				
(Teng Oil Plantation - 17°33'S 145°56'E)	930	2851 (1918-69)	2263 (1965-80)	79.38
Russel River				
(Bucklands - 17°23'S 145°58'E)	22.5	3838 (1919-69)	3441 (1918-80)	89.66
Tully River				
(Koombooloomba - 17°50'S 145°36'E)	163	2610 (1919-69)	2157 (1949-64)	82.64

Source: Bonnell et al (1986) using information from Queensland Water Resources Commission 1980.

ments). Knowledge of the interaction between short term rainfall and the soil hydraulic properties is fundamental to the understanding of the storm runoff generation process. The maximum 6 min. storm rainfalls show a marked change in frequency over the year controlled by the seasonal change in the synoptic climatology, which in turn causes a change in the preferred flowpaths for storm runoff. Four seasons can be recognised and the preferred equivalent hourly intensity range for maximum 6 min. rainfall in major storms are:

Summer Monsoon (December to March)   70-170 mm hr^{-1}
Post-Monsoon (April to mid-June)   25- 80 mm hr^{-1}
Winter (mid-June to September) usually less than   20 mm hr^{-1}
Pre-Monsoon (October and November)   up to 110 mm hr^{-1}

In some years persistent atmospheric stability results in the virtual absence of the pre-monsoon season.

   c. Frequency - intensity - duration. When the frequency - intensity - duration is considered for the Babinda catchments (Fig. 2) using RRUMS (Littleboy *et al*, 1986), 0.1 hr. equivalent hourly intensities range from 95 to 247 mm hr^{-1} and even absolute hourly intensities lie between 46 and 104 mm hr^{-1}. These figures emphasise the high short term intensities which occur during the summer monsoon season. But what makes this area outstanding is the long duration of rainfall events in comparison with the equatorial rainforest areas as a consequence of the mentioned differences in synoptic meteorology. This is shown by the 24 hour amounts which range between 8 and 27.5 mm hr^{-1}.

   d. Mt Bellenden Ker. Discussion so far has considered rainfall typical of the lowland areas of the wet tropical coast. Further west on the mountain range, more impressive rainfall amounts are recorded. A rain gauge has been in operation on Bellenden Ker Top since 1972. Mean annual rainfall at this site is 7664 mm, 1974-84 inclusive (median 7210 mm, range 6305 mm (1974) to 11,346 mm (1977)). But as Hall (1984) noted, 'Bellenden Ker Top has set new records (for Australia) in January 1979 for the highest daily fall (1140 mm), the highest weekly fall (3847 mm), the highest monthly fall (5387 mm) and the highest yearly fall (11,346 mm, 1977)'. Continuous rainfall recorders have only recently been installed at this site by the Department of Geography, James Cook University, but records so far show the persistence of one-minute intensities ranging between 30 to 75 mm hr^{-1} during events of long duration (Bonell *et al*, 1986). It is from such mountain

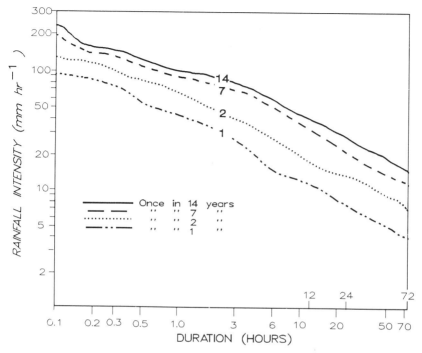

FIGURE 2. Rainfall intensity-frequency-duration analysis for the Babinda experimental catchments (1971-1983).

areas that make a major contribution to the recurrent disastrous flooding in the lower reaches of the principal rivers (Bonell, 1983).

III.   THE STORM RUNOFF HYDROLOGY

      Thus the wet tropical coast has to cope with frequent high intensity, long duration events in the summer months when most of the annual rainfall occurs. As part of a long term experiment between the Queensland Department of Forestry and the Department of Geography, James Cook University, the runoff generation process was investigated in the Babinda catchments occupying lowland tropical rainforest. The results of that work have already been reviewed in some detail (Gilmour *et al*, 1980; 1982; Bonell *et al*, 1981; 1986; 1987) and thus only a brief summary is given here.

A. *Tropical Rainforest - the Babinda South Creek*
    Experimental Catchment

Hydrological process work was concentrated in the
undisturbed South Creek catchment (25.7 ha) which is under-
lain by kaolin-dominated silty clay loam to clay loam soils
that may be up to 6 m in depth (red podzolic, Stace *et al*,
1968; Bonell *et al*, 1981; 1983) formed from basic metamorphic rocks, the Babalangee Amphibiolite of de Keyser
(1964). Previous water balance work by Gilmour (1975) showed
that about 65 percent (2515 mm) of the annual rainfall
(4009 mm) appears as runoff. About 47 percent of the South
Creek annual discharge appears as quickflow, the bulk of
which occurs in the six month period, December to May. More
significant is that frequently well in excess of 45 percent
of gross rainfall for individual storms appears as quick
flow during the wet season, with percentages exceeding 70
percent not unusual in the heaviest storms. Under such
conditions, instantaneous peak discharges greater than
5000 Ls^{-1} (70 mm hr^{-1}) can occur during monsoon storms
(January to March, 1976 and 1977) (Bonell *et al*, 1980). The
average lag response times between rainfall-stream discharge
is 0.4 hrs (mode 0.3 hrs, range 0.2-0.9 hr) (Bonell *et al*,
1986).

In terms of peak runoff volumes and lag times between
rainfall and stream discharge, this catchment has a response
more akin to an environment where Hortonian overland flow
(as defined by Kirkby, 1978, p.368) is the dominant
contributor to stormflow rather than subsurface stormflow
(as defined by Kirkby, 1978, p.373) generally associated
with humid forested lands especially in temperate areas.
For example, Dunne (1978, Figures 7.18 and 7.19) estimated
a peak runoff rate of 0.6 mm hr^{-1} and a lag time of 12.99
hr for a similar catchment area of 0.257 km^2. These diagrams
were a condensation of results from humid temperate catchments (some of which were forested) where subsurface
stormflow is claimed to be the dominant contributor to
stormflow. In contrast Dunne's (1978, Figures 7.7 and 7.8)
other diagrams show that for the same sized catchment peak
runoff rates of 150 mm hr^{-1} with lag times of 0.32 hr can
be expected where Hortonian overland flow dominates (Bonell
et al, 1987).

Detailed process work (Gilmour *et al*, 1980; Bonell *et
al*, 1981; 1987) showed that saturation overland flow (as
defined by Kirkby, 1978, p.371) was the dominant contributor
to streamflow in the summer monsoon season. As rainfall
intensities decline in the post-monsoon, subsurface storm-

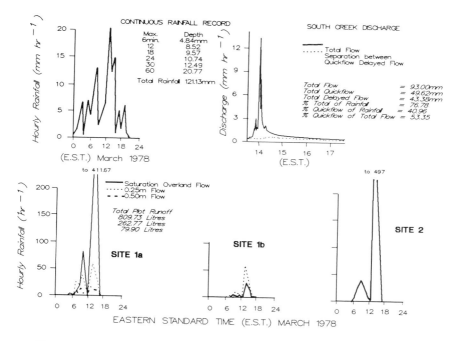

FIGURE 3. *The continuous record for rainfall, saturation overland flow, subsurface flow and South Creek discharge resulting from a storm on 14 March 1978, 00.00-19.00 h Eastern Standard Time.*

(a) No reliable record for subsurface flow at site 2 but earlier data collected indicated a significant response below 0.25 m. Total 0.25 m flow for individual storms was normally an order of magnitude lower than the total for saturation overland flow (Bonell and Gilmour, 1978).

(b) Plot areas: site 1a (0.023 ha), site 1b (0.006 ha), site 2 (0.016 ha). To avoid redirecting any major flow paths, the runoff plots were not bounded, but remained open to receipt of all runoff upslope. Consequently, the exact area contributing to runoff is not known and the figures quoted are only approximate (Source: Bonell et al, 1986).

flow from the root layer (< 0.25 m depth) of the rainforest soils became a more significant flow pathway in some parts of the catchment. During the winter when SE stream showers prevail, the hillslope runoff response by comparison was innocuous.

Figure 3 shows the various responses to a wet season storm at three runoff sites using two metre troughs located

at the surface, 0.25 m and 0.50 m. More experimental
details have been described elsewhere (Bonell *et al*, 1981;
Gilmour *et al*, 1980). The outstanding feature is the
occurrence of saturation overland flow at all sites despite
the short-term rainfalls for this particular below average
storm. Furthermore, there is a close relationship between
saturation overland flow and 0.25 m subsurface flow graphs
and the quickflow component of the hydrograph, thus highlighting the importance of saturation overland flow as a
major contributor to storm runoff.

This is an unusual storm runoff response for a forested
environment and can be attributed to several factors:

(1) Despite the highly transmissive surface soil layers
with log mean field saturated hydraulic conductivities,
K of > 800 mm hr^{-1} for the 0-0.1 m layer and 55 mm hr^{-1} for
the 0.1-0.2 m layer, the prevailing high rainfall
intensities frequently exceed the subsoil K (0.2-0.5 m
layer) which have a catchment log mean of 1.4 mm hr^{-1} (n
= 219).

(2) Antecedent soil moisture is high during the wet
season due to the frequency of rain events and can range
up to 1.2 to 1.5 m equivalent depth of rainfall stored in
the top 3 m of soil. The corresponding soil water pressure
(matric potential) is only marginally below atmospheric
pressure (matric potential, 0 to -1 m) which means that
during the early part of storms, positive matric potentials
and therefore saturation and runoff quickly redevelop
throughout the basin soils with the onset of intense storms
(Bonell *et al*, 1981; 1983). A similar process was
described elsewhere (O'Brien, 1982; Abdul and Gillham, 1984;
Gillham, 1984) where the tension saturated zone (Freeze
and Cherry, 1979) of the capillary fringe is quickly converted to phreatic water having a positive pressure
potential.

(3) The average catchment slope of 19° ensures the
rapid routing of storm runoff out of the catchment.

(4) The high dendritic drainage density (0.23 m m^{-1})
is able to tap most of the surface runoff.

(5) The sparsely littered forest floor and absence
of herbaceous cover in comparison with temperate forests
reduces the surface hydraulic resistance to saturation overland flow.

(6) The large volume of biopores in the topsoil enables
the rapid penetration of rainwater to the subsoil layer
(> 0.2 m depth) causing the development of a perched water
table and saturation overland flow.

The frequent occurrence of widespread saturation overland flow and subsurface storm flow in this rainforest catchment, particularly in the summer monsoon, represents part of the extreme 'wet' hydrological situation in the context of Hewlett's variable source area concept (Hewlett, 1961; Hewlett and Hibbert, 1967). The proportion of South Creek contributing to quickflow will depend on the transit flux and distance saturation overland flow has to take before it is tapped by organised drainage lines, either perennial or ephemeral. The soil hydraulic properties, the intensity and long duration of many monsoon storms associated with circular tropical disturbances, the high drainage density and steep catchment slopes; all favour a very large proportion of South Creek frequently contributing to quick-flow. Any temporary decline in the duration and intensity of summer monsoon storms, or more generally in the post-monsoon season, allows the spatial variability of the subsoil field saturated hydraulic conductivity to control differences in the dominant stormflow over South Creek. At such times only parts of South Creek may contribute to stormflow. Preliminary results (in conjunction with D. Cassells) from topography-wetness modelling (O'Loughlin, 1986) also indicate support for this runoff pattern. It is significant that there is pedological evidence for natural erosion by saturation overland flow on the steeper slopes with only remnants of the A2 and A3 horizons surviving (Murtha , 1981, pers. comm.).

The conclusion from this rainforest study is in marked contrast with the few other studies undertaken in the humid tropics. Saturation overland flow appears to be unimportant except in highly localised areas (Douglas and Spencer, 1985; Lal, 1981; Lundgren, 1980; Nortcliff and Thornes, 1981; Walsh, 1980) based on work mostly in equatorial rainforest. In addition, the proportion of annual rainfall being routed out as total runoff from equatorial rainforest studies is a lot lower. Low and Peh (1985, pers. comm.) report runoff yields (quickflow and delayed flow) of only 9.1 and 15.4 percent from paired catchments near Kuala Lumpur with annual rainfalls of 1766 mm and 1833 mm respectively. Elsewhere Salati and Vose (1984) report only 19 to 26 percent of annual rainfall (about 2000 mm) appearing as total runoff from Amazon rainforest basins near Manaus.

B. Extrapolation of the Babinda Catchment Study to Other Rainforest Areas of the Wet Tropical Coast

By their very nature, detailed catchment studies are

site specific and the potential for extrapolating the conclusions from this work has to be addressed.   The key
factors to an understanding of the storm runoff process
are the spatial variation of rainfall intensities and soil
hydraulic properties.   Implicit in this discussion is that
regional rainfall intensities are similar to those of the
Babinda catchments.   This aspect is still under
investigation.   However, preliminary measurements of
regional patterns of soil hydraulic properties (Bonell *et
al*, 1983) are encouraging in that the runoff plots of the
Babinda catchments are part of the lower and higher end
of the K range.   On some soil types, eg yellow podzolics,
saturation overland flow will develop even more frequently.
     When the runoff response of the larger catchments of
the wet tropical coast are considered from published data
by the Queensland Water Resources Commission (1980) (Table
I), the annual runoff coefficients are similar or even
higher than those for the Babinda catchments.   Whilst the
estimates of average drainage basin rainfall can be
questioned because of the sparsity of data from the
mountainous areas, the proportion of gross rainfall as
runoff ranges between 58 and 90 percent.

C.   Sugar Cane Lands

     Preliminary results from work undertaken by the
Queensland Department of Primary Industries (DPI) indicate
that ground cover by sugar cane trash, as well as soil type,
significantly alters time to commencement of runoff and
runoff rates based on a rotating-disc rainfall simulator
(Prove *et al*, 1986) (Table II).   Under natural rainfall
however, monsoon storms are still capable of producing
large volumes of runoff from krasnozem soils despite their
superior hydraulic properties compared with the red podzolic.
Prove (1985, pers. comm.) reported that under various
cropping treatments on a krasnozem near Innisfail, the
proportion of gross rainfall appearing as storm runoff from
approximately 400 m hillslope strips was in the range
of those measured for the Babinda catchments.   For example,
the percentages of storm runoff for an event of 210 mm rain
in March 1985 were up to 72 percent.   Such runoff can be
either saturation overland or Hortonian overland flow or
a combination of both depending on the agricultural treatment.   Also the response times between rainfall and storm
runoff were short, less than 6 minutes for all treatments.
Similar response times between rainfall - saturation overland
flow were measured in the Babinda catchments (Bonell *et
al*, 1979; 1981).

Work is currently underway by the DPI, in conjunction with the Department of Geography, James Cook University, investigating the runoff process on this site in more detail. Present indications however are that it is not too dissimilar from the description of the Babinda catchments when referring to the conventional cultivation treatment (Prove, 1987, pers. comm.).

IV. HYDROLOGICAL IMPLICATIONS FOR LAND MANAGEMENT

The high rainfall intensities, long duration of storms and resulting frequent occurrence of either saturation overland flow or, on disturbed, compacted soils, Hortonian overland flow has major implications for land management practices in forestry and agriculture.

TABLE II. *Effects of Trash Cover on Surface Runoff and Sediment Concentration under Simulated Rainfall*

	Percent trash cover					
	Red podzolic			Krasnozem		
	0	50	100	0	50	100
Rainfall intensity $(mm\ hr^{-1})$	73	77	74	83	87	84
Runoff initiation time (minutes)	2.7	4.1	13.3	8.0	17.3	22.5
Rainfall before runoff (mm)	3	5	14	11	25	27
Runoff 5 minutes after rain (%)*	90	70	0	0	0	0
Runoff 30 minutes after rain (%)*	100	100	89	73	35	12
Sediment concentration gL^{-1}	48.0	36.0	7.3	9.5	7.1	2.1

*Runoff as percentage rainfall
Source: Prove et al (1986)

A.   Forestry

   1.   The Impact of Logging and Clearing of Tropical
         Rainforest on Storm Runoff Hydrology and Erosion.
It became apparent by the 1960s that unconstrained logging
was causing spectacular gully erosion up to 12 m deep
(Gilmour, 1977a), especially on logging or 'snig' tracks
This instigated two separate investigations within the
Freshwater Creek basin near Cairns (Gilmour, 1971) and in
the Babinda catchments (Gilmour, 1977b).
   The Babinda work followed the classic approach of
calibration of undisturbed, paired catchments (North and
South Creek), followed by monitoring the impact of logging
and later clearing of North Creek. Statistical details
of the work were presented by Gilmour (1977b) and further
interpreted in Gilmour *et al* (1982) in the light of process
studies concerning runoff generation. Particularly
outstanding was no detectable change in quickflow volume,
quickflow duration or time to peak after logging and clearing
in North Creek. In addition there was only weak
statistical evidence for a small increase in peak discharge.
These characteristics imply that there were only minor
changes in runoff, in terms of process and source areas.
   Major changes however, were evident in the water quality
of North Creek after the change in land use. Peak suspended
sediment concentrations rose from 180 mg L^{-1} before logging
to about 520 mg L^{-1} in the two years after logging.
Clearing brought a much more dramatic change in suspended
sediment concentrations in North Creek reaching values as
high as 4000 mg L^{-1} despite dilution from peak discharges.
From the data on stream sediment levels in North Creek,
Capelin and Prove (1983) estimated that annual suspended
levels rose from 4.8 tonnes ha^{-1} before disturbance to 10.9
tonnes ha^{-1} after logging and 59.6 tonnes ha^{-1} after
clearing. Bedload was not measured, but Douglas (1967)
suggested this factor may be as high as 50 percent of the
suspended load which raises the total loss to about 90
tonnes ha^{-1}.
   When the erodibility of various rainforest soil types
is considered, the quoted erosion losses from the Babinda
catchments are likely to be conservative if the same
activity had been undertaken on other catchments of the
same morphometry. Using the dispersion index of Middleton
(1930), the most stable soils are those derived from basalt,
basic metamorphic (Babinda catchments), colluvium and some
of the granite members (Bonell *et al*, 1986). The remaining
soils derived from granite and acid metamorphics were
erodible below the surface layer where organic matter is

not incorporated. Exposure of these subsoils by log haulage causes them to disperse easily from raindrop impact and concentrated overland flow especially on logging tracks. Thus overland flow can be a mixture of both saturation overland flow and Hortonian overland flow due to soil compaction.

Apparent from this Babinda study, is that the amount of vegetation and the conditions of the top 0.2 m of soil are relatively unimportant in terms of quickflow generation. But the same two factors are of great importance in controlling erosion. The frequency of widespread saturation overland flow in undisturbed rainforest due to the prevailing rainfall intensities and shallow impeding subsoil, ensures little change in the storm runoff hydrology even after varying degrees of soil compaction from logging and clearing.

The need for the development of watershed management controls was initially advanced by the Freshwater Creek study. Gilmour (1971) presented evidence to show that the principal sources of sediment in the creek were from poorly located undrained roads and 'snig' tracks, and from earth and log filled crossings. In the light of these findings, several guidelines were put into operation and included:

(1) Snigging and hauling through streams was prohibited.
(2) The use of streamside buffer strips at least 20 m from the stream were adopted to reduce the supply of sediment from 'snig' tracks, roads and logging ramps. These strips also ensured stream bank stability from the high discharges resulting from monsoon storms.
(3) Earth and log fill crossings were prohibited. The alternative was the construction of girder bridges.

Following the implementation of these guidelines, significant reductions in suspended load were evident. For example, the highest measured concentration was 188 mg L^{-1} after a 262 mm rainfall compared with c 780 mg L^{-1} after 66 mm storm before these initiatives (Gilmour, 1971).

Further refinements to Gilmour's management controls were subsequently developed in the light of understanding of regional hydrological processes developed in the Babinda Research Programme, and these have been routinely applied to all forest harvesting operations since 1981. The nature of these controls and the social and technical considerations were described in detail by Cassells *et al* (1984). Amongst the technical improvements were:

(1) The allotment of specifications for buffer strip

width based on watercourse width, soil type and catchment size.

(2) Maximum spacing of cross drains along 'snig' tracks and roads for different slope angles and soil erodibilities to reduce concentration of overland flow.

(3) Design and location of haulage roads, 'snig' tracks, stream crossings, landings and the general logging direction and methodology.

2.   *Exotic Coniferous Plantations*.   As part of a policy to achieve net self-sufficiency in forest products in Queensland, the development of exotic coniferous plantations (*Pinus caribaea* var *hondurensis*) has proceeded on the northern and southern fringes of the wet tropical coast near Kuranda on the elevated hinterland, west of Cairns and in the Ingham-Cardwell area.   These plantations are just beyond the tropical rainforest area on land formerly occupied by eucalypt woodland (species description in Cassells *et al*, 1982) and gully rainforest.

The impact of this plantation forestry on stream water quality at Cardwell (annual rainfall, 2190 mm) and Kuranda (annual rainfall, 2060 mm) was reported by Cassells *et al* (1982).   Comparisons were made between undisturbed woodland and plantations of different ages which incorporated various site preparation techniques, eg mounding, no buffer strips, undisturbed buffer strips, contour banks, grassed waterways. Although annual rainfall at the two sites examined is only 50 percent of the Babinda catchments, the concentration of this rainfall into the December to March period remains the same.   Therefore it is reasonable to infer that the runoff generation process is similar to that described for the tropical rainforest during the summer monsoon season. Thus the erosion potential therefore remains high.

The following conclusions emerged from this work:

(1) All catchments where the previous vegetation was cleared and the land ploughed as part of the site preparation showed the expected rapid increase in suspended load in streams.   This was particularly apparent at Kuranda where a maximum concentration of 622,165 mg L^{-1} was recorded as against a background range of 55-400 mg L^{-1} in the undisturbed catchment.   Extensive gullying and sheet erosion from widespread overland flow occurred even where preparation included contour ploughing, graded contour mounding and a network of grassed waterways.

(2) Plantations established without the conventional cultivation (no tillage) showed no measurable impact on stream water quality.

(3) Those plantations established using broadcast contour ploughing including contour banks to divert down-slope runoff into grassed waterways and streamside buffer strips, returned to pre-clearing stream suspended load levels within 2 to 3 years of establishment.

(4) The older plantations where no buffer strip protection had been applied still showed suspended load concentrations much higher than the undisturbed catchments. For example, at Cardwell a plantation established in 1969/70 had mean concentrations for the three years 1980-82, incl. ranging from 50 to 460 mg L^{-1} compared with 27-33 mg L^{-1} for the undisturbed catchment. It was concluded (Cassells et al, 1982) that the absence of the protective riverine buffer strip was the cause of these higher levels in the old plantations. Whilst these catchments had a good ground cover, the earlier clearing and site preparation caused the stream banks to remain unstable at a number of points.

This study indicated that plantation establishment does have an impact on stream sedimentation, but with application of management controls eg buffer strips, the impact rapidly declines after a few years. Furthermore the long cropping period of plantation forests offers more protection from erosion than intensive short rotation cropping, eg sugar cane, especially on unstable soils.

B. Agricultural Areas

1. Sugar Cane Lands. Early deforestation of tropical rainforest commenced after 1873 and by 1879 sugar cane growing had commenced in the Innisfail district. Subsequently large tracts of land were cleared for agricultural development. By 1981 the area set aside for sugar cane was 125,442 ha, beef pasture, 110,000 ha and horticulture (principally bananas) 2650 ha on the coast. On the Atherton Tableland in order of importance maize, peanuts, rice and potatoes occupy a total of 16,600 ha, other horticulture (1600 ha) and 75,000 ha are almost evenly divided for beef and dairy cattle (Capelin, pers. comm). Evident from these land use statistics is the significance of sugar cane lands. Furthermore the intensive rotation cropping of this commodity results in a high erosion potential from such areas because of the persistence of widespread overland flow. Sugar cane land occupies most of the favourable slopes. More recently however there have been attempts to plant sugar cane in rainforest areas, away from the floodplain of the major rivers to prevent frequent inundations, for example the Russell

River near Babinda.   This measure has increased the rate
of land degradation (Gilmour *et al*, 1980).

Surveys by the DPI (Wilson *et al*, 1982; Dawson *et al*,
1983) in the early 1980s alerted to the seriousness of caneland erosion.   The widely-used conventional cultivation
(CC) involves burnt cane harvest trash raked and burnt,
several tillage operations to control weeds, sub-surface
fertilising and no grassed cover.   Furthermore to facilitate
cultivation and mechanical harvesting, cane rows are aligned
in sympathy with slopes rather than on a contour layout.

The disastrous effects of this practice were highlighted
in Capelin's (1981, pers. comm.) survey near Innisfail in
the aftermath of 2260 mm of rain in the period 1-14 January,
1981.   Erosion losses were estimated to range from 10 tonnes
ha^{-1} on areas protected by plant residues to over 500 tonnes
ha^{-1} on bare fields cultivated on krasnozems, with slopes
up to 20 percent.   Spectacular gully erosion occurred up
to 3 m deep and 10 m wide.   A later survey by Capelin and
Prove (1983) estimated average annual soil losses of 100
tonnes ha^{-1} from 11,000 ha of cane lands on various soils
north of Cardwell.   These figures emphasise the severity
of erosion from even the least erodible soils eg krasnozems,
when cultivation takes place on steep slopes.

One of the problems with the CC technique concerns lack
of protection of the bare soil surface from harvest until
February or March when the full crop canopy will provide
some crop protection (see monthly percentage crop changes
in Table V of Prove *et al*, 1986).   As a soil conservation
measure, a series of trash management trials were initiated
at five different sites near Innisfail on the wet tropical
coast by the Bureau of Sugar Experimental Stations (BSES)
and the DPI.   These incorported red podzolic, krasnozem
and red earth (Stace *et al*, 1968) soils on slopes ranging
from 5-16 percent (BSES/DPI, 1984).   The work was
subsequently extended (Prove *et al*, 1986).   The following
treatments were applied:

(1) conventional cultivation (CC)
(2) green cane harvest trash blanket (GCTB) and chemical
weed control if necessary
(3) burnt cane harvest trash blanket (BCTB) with
chemical weed control
(4) burnt cane harvest, trash removed, zero till (0
TIL) and chemical weed control.

Prove *et al* (1986) noted that green cane harvests
provided complete ground cover at most sites whilst burnt
cane trash blanket gave between 40 and 60 percent cover.

Furthermore only a small reduction (2-3 percent) of covers was detected in both GCTB or BCTB five months after harvest.

Although runoff volumes can be similar to the CC treatment, runoff velocities are commonly lower when trash cover is present (Prove *et al*, 1986; 1987, pers. comm). But the fact there was no statistically significant difference in soil movement between the GCTB, BCTB and O TIL treatments suggests that zero or minimum tillage as well as trash retention is the major soil conservation factor reducing erosion (Prove *et al*, 1986). Losses from these three treatments were approximately 20 tonnes ha^{-1} for the 1982-83 wet season as against 75 to 150 tonnes ha^{-1} for the CC treatment (BSES/DPI, 1984).

These trials were undertaken using the traditional layout of cane rows up and down slope. With the implementation of further soil conservation and runoff control structures (eg across slope row direction, diversion banks, contour banks, waterways), even further reductions in soil loss might be attained. Prove *et al* (1986) put forward a series of soil conservation strategies which included in various combinations trash cover retention, minimum or no tillage, early planting and soil erosion/runoff control structures for different soil types and slopes.

2. River Aggradation. Despite the high erosion rates from sugar cane land, the fate of this eroded material has until very recently never been considered. A study was commissioned for Cameron McNamara Consultants to study the implications of soil loss in the drainage basin of the South Johnstone river (530 km^2) near Innisfail. The results are reported in detail by Cameron McNamara (1985) and summarized by Connor (1986).

River aggradation over 40 years was assessed by the reduction of wet area in the lower 25 km of river using a digitising process on aerial photography for the years 1942, 1957, 1977 and 1983. This section incorporated drainage from 170 km^2, containing in 1983 some 97 km^2 of assigned cane land (Fig. 4). This reach of the river was divided up into nine different zones for final comparison.

The percentage reduction in wet area for the zones is shown in Figure 4. Accumulation had occurred in the river channel at a reasonably steady rate from 1942 to 1977 with a total wet area reduction of 167,180 m^2 for zones 2 to 9 combined giving an accretion rate of 4780 m^2 per annum (Connor, 1986). Between 1977 and 1983 an additional 128,640 m^2 of wet area was lost, raising the rate to 21,440 m^2 per annum. If this rate is valid and not a temporary fluctuation due to a flood event for example, then Connor

(1986) suggested this increase could be the result of sugar cane expansion into highly erosion prone areas of steep basalt and metamorphic soils. About 5490 ha of 9700 ha of cane land in the South Johnstone catchment was in such areas in 1983.

Modelling (Cameron McNamara, 1985, pp.19-21; Connor, 1986) indicated flood stages at some points could rise by up to 1.3 m for the 1 in 10 year and 1 in 50 year floods in 1983, compared with 1942, as a result of this channel aggradation.

3. *Pasture.* No published material is presently available on the effects of storm runoff-erosion processes

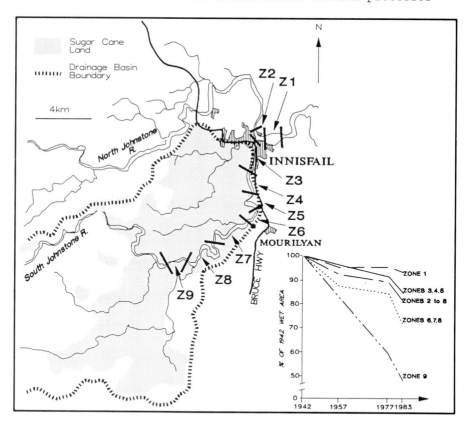

FIGURE 4.   *The historical reduction in wet area for various combinations of reach of the lower South Johnstone River.   The 9 zones of reach are shown in the inset and the boundaries of wet plan area are the edge of the vegetated banks (Source: Cameron McNamara, 1985).*

on pasture, but it is evident from field observations that under well managed pasture, the soil is effectively shielded from raindrop impact and there are many impediments to surface wash. In these situations there is little evidence of severe soil erosion after intense summer monsoon storms.

V. CONCLUSION

The geographical position of the wet tropical coast of north-east Queensland in relation to the 'maritime continent' and associated inter-hemisphere exchange of air makes it one of the most meteorologically active areas of the tropics. The nature of the monsoon trough causes an alternation of intense bursts of rain, which can last for several days, followed by periods of fine weather. Few other areas of the world have a comparable synoptic climatology with the possible exception of the north-east Bay of Bengal area in the summer monsoon and Vietnam (September/October) (Sadler, 1985, pers. comm.).

This area occupies part of the extreme wet spectrum in comparison with other tropical rainforest areas and ensures that widespread saturation overland flow is common through the interaction of the prevailing rainfall with the soil hydraulic properties. A high proportion of annual as well as individual storm inputs are quickly routed out as quickflow. This is an unusual runoff response for undisturbed forest in view of the high surface infiltration capacities and emphasises the close links between storm drainage and synoptic climatology. Such links have largely been ignored elsewhere in runoff process studies because most research has been undertaken in the humid temperate areas, such as the eastern United States or western Europe, where rainfall intensities are much lower in magnitude (eg Hewlett *et al*, 1977; Weyman, 1973).

In a review of runoff processes and models in the humid tropics, Walsh (1980, p.181) noted that of all the tropical rainforest areas so far investigated, the Babinda catchments have 'the only *distinctively tropical* runoff process pattern'. Clearly the high rainfall intensities and long duration of storms outweigh the lithologic influences in the Babinda catchments. Otherwise the soil hydraulic properties would have suggested a localised saturation overland flow and subsurface stormflow model found elsewhere in humid tropical and temperate areas (Kirkby, 1978; Walsh, 1980).

The frequency of widespread saturation overland flow

ensures natural erosion occurs on the steeper slopes of
undisturbed forest and poses severe land degradation problems
from forestry practices and on conversion to agriculture,
especially frequent rotation crops such as sugar cane.   The
high prevailing rainfall ensures that the amount of
vegetation and condition of the top 0.2 m of soil are
relatively unimportant in quickflow generation.   But the
same two factors are of great importance in controlling
erosion.   After decades of protracted soil loss, it was
only in the 1980s that long overdue land management controls
were initiated in both forestry and agriculture.   Ironically
this is being undertaken at a time when the socio-economics
of both activities are coming under scrutiny.   Dwindling
supplies of good rainforest timber coupled with increasing
pressure from conservationists might cause rainforest
harvesting to cease. In addition the collapse of sugar prices
in the mid-1980s, if protracted, might reduce the area under
sugar in the long term.   For the remaining areas, present
economics might accelerate the adoption of soil conservation
measures by farmers.   Estimates in 1983 (Mackson, 1983)
indicated tillage and weed control costs per hectare could
be reduced by A$172 and A$69.50 for green cane trash blanket
and burnt cane trash blanket respectively, compared with
conventional cultivation.   The urgency for the implementation
of these measures is evident elsewhere from the
deterioration in river water quality and channel morphology.

On the other hand, the wet tropical coast of north-east
Queensland in common with other tropical areas of Australia
offers unique advantages over 'expeditionary' tropical
research operating out of North America and Western Europe.
The political stability, availability of western
technology and accessibility of this area ensures the
ability to conduct *long term* hydrology-erosion studies.
This aspect combined with the severity of the wet season
means that the wet tropical coast is an ideal laboratory
for developing solutions to land management issues.   Such
findings have great potential for extension to other
developing areas of the humid tropics.

REFERENCES

Abdul, A. and Gillham, R.W. (1984)   Laboratory studies of the
    effects of the capillary fringe on stream flow
    generation. *Water Resour. Res.*, 20, 691-698.
Bonell, M.   (1983)   Hydrology of Humid Tropical Regions:
    Wet Tropical Coast of Northeast Queensland.   Report for
    UNESCO IHP Project A.1.10.

Bonell, M. and Gilmour, D.A. (1978) The development of overland flow in a tropical rainforest catchment. *J. Hydrol.*, 39, 365-382.

Bonell, M. and Gilmour, D.A. (1980) Variations in short term rainfall intensity in relation to synoptic climatological aspects of the humid tropical north-east Queenlsand coast. *J. Trop. Geog.*, 1, 16-30.

Bonell, M., Cassells, D.S. and Gilmour, D.A. (1987) Spatial variations in soil hydraulic properties under tropical rainforest in north-eastern Australia. In Yu-si Fok (ed.) *Int. Conf. on Infiltration, Development and Application*, Univ. of Hawaii, 153-165.

Bonell, M., Gilmour, D.A. and Sinclair, D.F. (1979) A statistical method for modeling the fate of rainfall in a tropical rainforest catchment. *J. Hydrol.*, 42, 241-257.

Bonell, M., Gilmour, D.A. and Sinclair, D.F. (1981) Soil hydraulic properties and their effect on surface and subsurface water transfer in a tropical rainforest catchment. *Hydrol. Sci. Bull.*, 26, 1-18.

Bonell, M., Gilmour, D.A. and Cassells, D.S. (1983) A preliminary survey of the hydraulic properties of rain-forest soils in tropical northeast Queensland and their implications for the runoff process. *Catena Supplement* No. 4, 57-78.

Bonell, M., Gilmour, D.S. and Cassells, D.S. (1986) The storm runoff response to various rainfall systems on the wet tropical coast of northeast Queensland. East-West Centre, Environ. and Policy Inst. Working Paper, Honolulu, Hawaii.

Bureau of Meteorology (1978) *Australian Tropical Cyclone Forecasting Manual*. Aust. Bur. of Meteor.

Bureau of Sugar Experimental Stations Qld Dept. Prim. Ind. (1984) A review of results of trails with trash management for soil conservation. *Proc. Aust. Soc. Sugar Cane Technol. Conf.*, 101-106.

Cameron McNamara Consultants (1985) *Stream Aggradation Effects of Soil Erosion*, Brisbane.

Capelin, M.A. and Prove, B.G. (1983) Soil conservation problems of the humid coastal tropics of north Queensland". *Proc. Aust. Soc. Sugar Cane Technol. Conf.*, 87-93.

Cassells, D.S. and Gilmour, D.A. (1978) The changing role of the north Queensland rainforest. *Proc. 8th World Forestry Congress*, Jakarta.

Cassells, D.S., Gilmour, D.A. and Gordon, P. (1982) The impact of plantation forestry on stream sedimentation in tropical and sub-tropical Queensland - an initial assessment. *Agricultural Engineering Conference*, Armidale, Inst. of Engrs. Aust.

Cassells, D.S., Gilmour, D.A. and Bonell, M. (1984) Watershed forest management practices in the tropical rainforests of north-eastern Australia. Proc. IUFRO Symp. on Effects of Forest Land Use on Erosion and Slope Stability. *N.Z. For. Res. Inst. Publ*, 289-298.

Connor, T.B. (1986) The impact and economic effects of soil conservation practices on river aggradation. IAHS Publ. No. 159, 81-91.

Davidson, N.E., McBridge, J.L. and McAvaney, B.J. (1983) The onset of the Australian monsoon during Winter MONEX: Synoptic Aspects. *Monthly Weath. Rev.*, 111, 496-516.

Dawson, N., Berndt, R. and Venz, B. (1983) Land Use Planning - Queensland Canelands. *Proc. Aust. Soc. Sugar Cane Technol. Conf.*, 43-52.

de Keyser, F. (1964) Explanatory Notes on the Innisfail, Qld. Sheet SE/55-6. Bur. Min. Res., Geol. and Geophys. Australia.

Douglas, I. (1967) Natural and man-made erosion in the humid tropics of Australia, Malaysia and Singapore. *IAHS, Publ.* 75, 17-30.

Douglas, I. and Spencer, T. (1985) Present-day processes as a key to the effects of environmental change. In I. Douglas and T. Spencer (eds.) *Environmental Change and Tropical Geomorphology*, Allen and Unwin, London, 39-73.

Dunne, T. (1978) Field studies of hillslope processes. In M.J. Kirkby (ed.) *Hillslope Hydrology*, Wiley, Chichester, 227-293.

Freeze, R.A. and Cherry, C. (1979) *Groundwater*, Prentice-Hall, New Jersey.

Gillham, R.W. (1984) The capillary fringe and its effect on water table response. *J. Hydrol.*, 67, 307-324.

Gilmour, D.A. (1971) The effects of logging on streamflow and sedimentation in a north Queensland rainforest catchment. *Commonwealth Forestry Review*, 50, 38-48.

Gilmour, D.A. (1975) Catchment water balance studies on the wet tropical coast of north Queensland. Unpublished PhD Thesis, James Cook University of North Queensland, Townsville.

Gilmour, D.A. (1977a) Effects of logging and clearing on water yield and quality in a high rainfall zone of north-east Queensland. *Inst. Engrs. Aust. Conf. Publ.*, 77/5, Canberra, 156-160.

Gilmour, D.A. (1977b) Logging and the environment, with particular reference to soil and stream protection in tropical rainforest situations. *FAO Conservation Guide 1*, Rome, FAO, 223-235.

Gilmour, D.A., Bonell, M and Sinclair, D.F. (1980) An investigation of storm drainage processes in a tropical rainforest catchment. *Aust. Water Res. Coun. Tech. Paper 56*, Canberra.

Gilmour, D.A., Cassells, D.S. and Bonell, M. (1982) Hydrological research in the tropical rainforests of north Queensland: some implications for land use management. *Inst. Engrs. Aust.*, 28.6, 145-152.

Hewlett, J.D. (1961) Watershed management. Report US Forest Service, Asherville, North Carolina.

Hewlett, J.D. and Hibbert, A.R. (1967) Factors affecting the response of small watersheds to precipitation in humid areas. *Proc. Int. Forest Hydrol.*, Penn. State Univ., Pergamon, 275-290.

Hewlett, J.K., Fortson, J.C. and Cunningham, G.B. (1977) The effect of rainfall intensity on stormflow and peak discharge from forest land. *Water Resour. Res.*, 13, 259-266.

Isbell, R.F., Webb, A.A. and Murtha, G.G. (1968) *Atlas of Australian Soils*. Sheet 7, North Queensland with explanatory data. CSIRO and Melb. Univ. Press.

Kirkby, M.J. (ed.) (1978) *Hillslope Hydrology*, Wiley, Chichester.

Lal, R. (1981) Deforestation of tropical rainforest and hydrological problems. In R. Lal and E.W. Russell (eds.), *Tropical Agricultural Hydrology - Watershed Management and Land Use*, Wiley, Chichester, 131-140.

Littleboy, M., Silburn, M. and Rosenthal, K. (1986) RRUMS Version 3.0 User Manual. Qld. Dept. Prim. Ind., Brisbane.

Lourensz, R.S. (1981) *Tropical cyclones in the Australian region, July 1909 to June 1980*. Aust. Govt. Publ. Service, Canberra.

Lundgren, L. (1980) Comparison of surface runoff and soil loss from runoff plots in forest and small scale agriculture in the Usambara Mts., Tanzania. *Geog. Annal.*, 62, 113-148.

McAlpine, J.R. and Keig, G. with Falls, R. (1983) *Climate of Papua New Guinea*. CSIRO in assoc. with ANU Press, 11-38.

Mackson, J. (1983) Trash retention: dollars in your pockets. *Aust. Canegrower*, 22-24.

Middleton, H.E. (1930) Properties of soils which influence soil erosion. *USDA Tech. Bull.* 178, 1-16.

Murtha, G.G. (1986) Soils of the Tully-Innisfail Area, North Queensland, *CSIRO Div. Soils*, Report No. 82.

Nicholson, D.I., Henry, N.B., Rudder, J. and Anderson, T.M. (1983)   Research basis of rainforest management in north Queensland.   Paper at XV Pacific Science Congress, Feb. 1983, Dunedin, New Zealand.

Nortcliff, S. and Thornes, J.B. (1981)   Seasonal variations in the hydrology of a small forested catchment near Manaus, Amazonas, and the implications for its management. In R. Lal and E.W. Russell (eds.), *Tropical Agricultural Hydrology - Watershed Management and Land Use*. Wiley, Chichester, 37-57.

O'Brien, A.L. (1982)   Rapid water table rise. *Water. Resour. Bull.*, 18, 713-715.

O'Loughlin, E.M. (1986)   A method for analysing catchment topography for application to hydrology and land conservation. *Water Resour. Res.*, 22, 794-

Prove, B.G., Truong, P.N. and Evans, D.S. (1986)   Strategies for controlling caneland erosion in the wet tropical coast of Queensland.   *Proc. Aust. Soc. Sugar Cane Technol. Conf.*, 77-84.

Queensland Water Resources Commission (QRRC) (1980) *Queensland Streamflow Records to 1797*, Vol. 1, QWRC,

Ramage, C.S. (1968)   Role of a tropical 'maritime continent' in the atmospheric circulation. *Monthly Weath. Rev.*, 96, 365-370.

Sadler, J.C. and Harris, B.E. (1970)   The mean tropospheric circulation and cloudiness over South-East Asia and neighbouring areas. *Hawaii Institute of Geophysics*, Univ. of Hawaii.

Salati, E. and Vose, P.B. (1984)   Amazon Basin: a system in equilibrium. *Science*, 225, 129-138.

Soil Survey Staff (1975) *Soil Taxomony, a basic system of soil classification for making and interpreting soil surveys*, US Dept. of Agric. Hdbk. No. 436.

Stace, H.C.T., Hubble, G.D., Brewer, R., Northcote, K.H., Sleeman, R. J., Mulcahy, M.J. and Hallsworth, E.G. (1968)   *A Handbook of Australian Soil*, Rellim Tech. Publ., Glenside, South Australia.

Sumner, G.M. and Bonell, M. (1986)   Circulation and daily rainfall in the North Queensland wet season 1979-1982. *J. Climat.*, 6, 531-549.

Walsh, R.P.D. (1980)   Runoff processes and models in the humid tropics. *Zeit. f. Geomorph. Suppl. Band,* 36, 176-202.

Weyman, D.R. (1973)   Measurements of the downslope flow of water in a soil.   *J. Hydrol.*, 20, 267-288.

Wilson, T.D., Wiseman, A.F. and Dwyer, G. (1982)   *Soil conservation practices and related attitudes of Innisfail canegrowers*. Qld. Dept. Prim. Ind., Project Rept. QO 82015.

4
Channel Sediment Loads: Comparisons and Estimation

W.A. Rieger
L.J. Olive

Department of Geography and Oceanography
University College
University of New South Wales
Australian Defence Force Academy
Campbell,   ACT

I. INTRODUCTION

Channel sediment load is the resultant of a complex set
of processes operating within a catchment area (Walling,
1984). There is differential erosion on the slopes,
depending on highly localised factors such as the nature
of slope material and slope angle. Some of this eroded
material is simply redistributed on the slopes while a
portion of it is transported to the stream channel via some
sediment delivery mechanism.

The term "sediment delivery ratio" (SDR) has been used
to conceptualise the relationship between slope erosion and
channel sediment loads in a catchment (Walling, 1984).   SDR
is simply the ratio of sediment carried by a stream to the
eroded sediment within a catchment.   In terms of SDR,
channel sediment loads can be considered as the sediment
yield from the catchment and is dependent on a sediment
delivery process which transports eroded slope sediment
to the channel.

This framework, where channel sediment loads are viewed
as sediment yields or as a measure of eroded material being
removed from a catchment, is important in that it considers
more than just the sediment loads. It incorporates both
the sediment generation on the slopes and the delivery of
sediment to the channel as well as the spatial, temporal

and scalar variations inherent in these two factors (see Pickup, this volume).

The following discussion is based on the sediment yield approach. A brief review of the general nature of sediment load and of its calculation is followed by a discussion of Australian sediment yields and denudation rates. The current debate on methods of estimating sediment yield is then outlined. Finally the temporal and spatial aspects of sediment yields are discussed.

II. GENERAL ASPECTS OF SEDIMENT LOAD

A. *The Relationship Between Load, Yield and Denudation*

The sediment loads transported within a stream channel can be divided into two broad categories: suspended load and bed load. Suspended load includes mineral and organic material which is supported by fluid turbulence. Bed load consists of sediment particles which are moving on or near the stream bed. The distinction between the two categories is rather arbitrary in that it depends on flow velocity and turbulence.

For the most part, research on sediment yields has been based on suspended load. There are a number of reasons for this approach. In most rivers, the majority of the total sediment load is made up of suspended load (Walling and Webb, 1983). Suspended load is easier to measure than bed load and simply involves the determination of suspended sediment concentration from water samples taken from the stream channel. Finally, the methodology for estimating suspended load from channel data is well established in geomorphology and relatively easy to use.

Suspended load is calculated from channel measurements of discharge ($m^3 sec^{-1}$) and suspended sediment concentration ($mg\ L^{-1}$). The expression:

$$L = k \int_{0}^{t} CQ\ dt \tag{1}$$

where C is concentration,
Q is discharge,
k is a conversion factor,

gives the suspended load, L, say in kg, for a the time interval 0 to t.

Suspended sediment yield is the standardisation of load to the average rate of material being removed from the catchment area (Chorley *et al*, 1984). It is usually expressed in t $km^{-2}y^{-1}$ and allows for the comparison of yields between catchments of varying sizes and for varying time intervals during which load has been measured.

Denudation is simply the conversion of the mass measure of sediment yield to a volume measure, or by the division of yield by the specific gravity of the eroded material. A value of 2.5, the specific gravity of rock, is commonly used for denudation calculations and assumes that the soil thickness remains constant throughout the catchment over long periods of time. Denudation is expressed in $m^3km^{-2}y^{-1}$ which is equivalent to an erosion rate in mm $(1000y)^{-1}$. Generally denudation is considered to be an erosion rate over a long period of time (Chorley *et al*, 1984).

Suspended sediment yield and denudation, then, are expressions which standardise channel sediment load. The standardisation procedure averages a point load, measured over a given time period to spatial measure for a fixed time interval. This allows comparisons of channel sediment loads for different sized catchments and for different periods of measurement of channel discharge and suspended sediment concentration.

B. *Methods of Estimating Channel Sediment Load*

The most commonly used method of determining channel sediment load is the sediment rating curve which establishes a fixed relationship between load and discharge for a particular catchment. The field data for generating the curve consist of paired values of discharge and concentration over a wide range of discharge. For each paired set of observations, load is calculated using a discrete form of Eqn. 1. The actual curve is developed with use of regression on the log-transformed values of concentration and discharge and takes the form:

$$L = aQ^b \qquad\qquad\qquad (2)$$

where *a* and *b* are the calculated regression coefficients.

The rating curve approach has appeal to researchers because of its ease of use. Once it has been established, for a particular catchment, load can be estimated from readily available discharge data. Mean hourly or mean daily discharge is calculated and simply substituted into Eqn. 2 to give estimates of suspended sediment load.

Another method for estimating channel sediment load involves the continuous or short interval monitoring of discharge and concentration and then calculating load directly using Eqn. 1. One form of continuous monitoring uses automatic water samplers which have the advantage of giving a more accurate estimate of load. However, the great expense and time involved with analysing the water samples for suspended sediment concentration, usually results in a sampling programme based on hourly sampling intervals. Such an interval may be appropriate for large rivers, but can miss the important short term variations in concentration and discharge in smaller rivers.

A second form of continuous monitoring uses turbidity meters to indireclty calculate suspended sediment concentration from turbidity. Given the appropriate data logging equipment, turbidity can be recorded at the same short time intervals as discharge, thus picking up the short term variations between the two variables. A difficulty with this method of calculating load is the possible changes in the concentration-turbidity relationship due to such factors as water colouration and particle size. A further problem includes the relationship between turbidity measured at a point and mean concentration in the whole section which varies with discharge. However, the load calculations based on a short time interval more than likely cancel out the error in estimating concentration from turbidity.

III. AUSTRALIAN SEDIMENT YIELDS

A. *Global Comparisons*

A number of authors have attempted to determine estimates of continental sediment yield (Holeman, 1968; and

Strakhov, 1967). Care should be used in interpreting their results for two major reasons. First, they draw on a large number of sources where different methodologies and sampling techniques were used in estimating sediment yield. Such differences can lead to variations in yield, as evidenced by a New Zealand case where two researchers used the same data set and their yield estimates differed by two orders of magnitude (Walling and Webb, 1983). Second, there is the problem of extrapolating small scale catchment yields to large scale continental estimates of yield. Although there are short comings, such exercises can give a broad view of Australian sediment yields.

Table 1 summarises the work of Holeman (1968) and Strakhov (1967) and the differences between the two authors gives some idea of the problems of estimating continental sediment yield. These differences aside, both agree that Australian sediment yields are amongst global minima.

Walling and Webb (1983) use a sub-continental scale for a more detailed analysis of global patterns of sediment yield. The highest yield are associated with the loess areas of China and with Pacific rim mountain areas. They estimate the average global sediment yield as 150 t $km^{-2}y^{-1}$ and the average Australian sediment yield as 100 t $km^{-2}y^{-1}$, well below the global average.

A number of possible causes have been postulated for Australia's low continental sediment yields. Because deep chemical weathering and size reduction to clay rich material is common to most of the continent, there is certainly a considerable supply of source material for transport. Walker and Hutka (1979) have found that the B horizons of podzolic soils are dominated by clay sized particles. However, the low gradients and large temporal variability in flow of major Australian rivers (McMahon, 1983), combine to give an inefficient sediment delivery system as detailed by Olive and Rieger (1986).

B. *Catchment Level Yields*

Until recently, studies on Australian sediment yields have been limited in either the duration of the monitoring period or to relatively small catchments. Most studies were the result of thesis research and involved the analysis of small coastal catchments over short periods of say, one to two years. Recently, however, more intensive and long term projects have been established to determine the impact of catchment disturbances such as logging, mining and urbanisation.

TABLE I: Continental Sediment Yields[a]

Continent	Area (km^2 x 10^6)	Sediment yield (t $km^{-2}yr^{-1}$) Holman	Strakhov
Europe	9.67	35	43
Asia	44.90	600	166
Africa	29.80	27	37
North and Central America	20.40	96	73
South America	18.00	63	93
Australia	7.96	45	32

a   *Figures taken from Holeman (1968) and Strakhov (1967)*

Major reviews of sediment yield studies have been carried out by Olive and Walker (1982) and Loughran (1984). Their results are summarised in Table II and the locations of the catchments corresponding to the yields in this table are shown in Figure 1. As with the continental sediment yields, care should be taken in interpreting yield values because of differences in sampling schemes and sediment yield estimation techniques used by various authors.

The sediment yields in Table II broadly confirm that Australian sediment yields are low by world standards. Most yields are less than 50 t $km^{-2}yr^{-1}$ and the exceptions can be explained by either catchment disturbance or precipitation characteristics. The yield for Deep Ck. is high due to intensive agricultural use at the catchment The Babinda River is situated in tropical north Queensland where the high annual precipitation and high rainfall intensities explain the large yields (see Bonell, this volume). The Ord River catchment has low annual precipitation (400-800 mm) but most of the precipitation occurs in high intensity storm events during the monsoonal "wet", giving rise to high sediment yields.

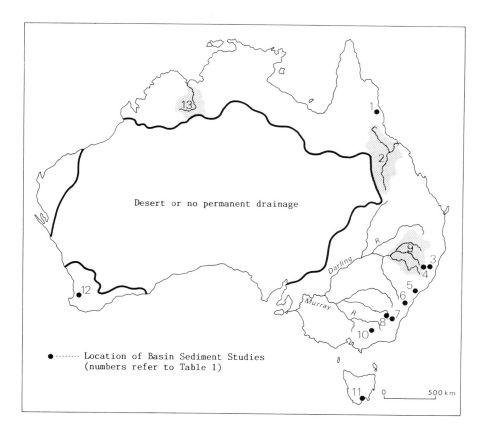

FIGURE 1. Location map.

The data in Table II also indicate a general inverse relationship between yield and catchment size, similar to that found by Schumm (1963). Sediment yields from small agricultural plots certainly strengthen this inverse relationship. Ciesolka and Freebairn (1982) estimate yields of up to 5300 t $km^{-2}y^{-1}$ on a one ha plot in the Darling Downs area of Queensland.

IV. ACCURACY OF RATING CURVE ESTIMATES

A. *Rating Curve Under Estimation*

As mentioned above, the rating curve methodology has appealed to researchers faced with the problem of estimating

TABLE II: Australian Basin Sediment Yields

Basin	Map locality (Fig. 1)	Area (km²)	Mean Ann Rain (mm)	Sediment Yield (t km^{-2}yr^{-1})	Source
QUEENSLAND					
Tully R	1 ⎫	44-585	1015-	8-45	Douglas (1973)
Herbert R	1 ⎬		2235		Douglas (1973)
Babinda	1	15	4320	480	Douglas (1983)
Burdekin R	2	129500	400-2900	23.17	Belperio (1979)
NEW SOUTH WALES					
Macleay R	3	7.5-20	1500-2000	138-179	Loughran (1969)
Chandler R	3	208	950	6.2-8.4	Loughran (1976)
Macleay R	4	<16	872-1066	.33-12.52	Field (1985)
Congewai Ck	5	85.5	1084	28	Loughran (1977)
Deep Ck	5	25	764	121-232	Geary (1981)
Macquarie Rivulet	6	1-7	1500	8-17	Douglas (1973)
Queanbeyan R	7	284	475	2.7	Douglas (1973)
Goodradigbee R	8	26-216	400-1000	1.5-11.7	Douglas (1973)
Barwon R	9	139000		2.9	Taylor (1976)
VICTORIA					
East Kiewa R	10	1.4-2.4	1850	9.9-19.5	Leitch (1982)
TASMANIA					
Browns R	11	13	680-1220	12	Olive (1973)
Mountain R	11	40	780	11	Olive (1973)
Snug Rivulet	11	19	1140	12	Olive (1973)
WESTERN AUST					
Wights R	12	0.93	1150	57	Abawi & Stokes (1982)
Ord R	13	46100	400-800	634	Kata (1978)

channel sediment loads. The appeal is based both on ease of use and cost. Once the curve has been calibrated for a given catchment, only discharge measurements are required for the load calculations. These discharge measurements are relatively easy to acquire and involve monitoring stage at a river cross section. In contrast, continuous monitoring techniques require field measurements of discharge and suspended sediment concentration. The latter is calculated either directly from water samples or indirectly from turbidity measurements, both of which involve added costs in terms of equipment and labour.

Though more expensive to use, the continuous monitoring method can provide data which can be used to assess the accuracy of rating curves. Because continuous monitoring samples both discharge and concentration over time, it provides a relatively accurate estimate of suspended sediment load. The smaller the sampling time interval, the more accurate is the load estimate.

Walling (1977; 1978) was one of the first researchers to carry out such a comparison between rating curve estimates and continuous monitoring estimates. Turbidity meters were used to indirectly measure concentration for river catchments located in south west England. Results showed that rating curves under estimated load by up to 80%. Similar findings have been reported by Olive et al (1980) for five small forested catchments near Eden, N.S.W. In this case, automatic water samplers were used to obtain suspended sediment concentrations.

Ferguson (1986) has suggested that the under estimation by rating curves is due to bias inherent in the log-transformed framework of the prediction methodology. He suggests a correction factor in the form of:

$$exp \ (S/2),$$

where S the variance of the residuals about the fitted line or rating curve. Rating curve estimates are then multiplied by this correction factor to give unbiased estimates of load.

Oddly enough Ferguson is the first geomorphologist to mention this inherent bias in the power function which is a widely used statistical technique in fluvial geomorphology. The likely reason for this is that the rating curve is one of the few applications of the power function to a prediction framework. The bulk of applications of log-log regression use the fitted curve as a blunt instrument to determine the broad relationship between variables, giving little concern to the accuracy of estimation. An example of such a use is the past work on channel geometry and catchment morphology (Melton, 1958).

B. *The Rating Curve Format*

A more likely reason for the lack of estimation precision with the rating curve is related to the nature of the dependent and independent variables which are used to generate the curve. Load is calculated from a discrete version of Eqn. 1 or as a function of discharge. Thus discharge appears on both sides of the rating curve and introduces spurious correlation between the variables (Grenney and Heyse, 1985).

In effect, the rating curve postulates a relationship between discharge and itself. With such a relationship, the original concentration data are manifested as noise or residuals about the curve. The authors have carried out an exercise to show this to be the case. A rating curve was first calculated using 100 paired field observations of suspended sediment concentration and discharge, with load being calculated as in Eqn. 1.   A second curve was then fitted using a randomly generated variable in the place of concentration. This new variable had a similar distribution to the original concentration data. The difference between the resulting curve and the original rating curve was statistically insignificant.

Though Ferguson has made a valuable contribution in pointing out the inherent bias in regression with log-transformed variables, the application of his correction factor to suspended sediment rating curves is of little use. Because of the spurious nature of the relationship between load and discharge, the bias correction is somewhat similar to what Klemes (1986) says:

> "...in hydrologic modelling we concentrate on refining the computation of various hydrologically irrelevant trivia, while evading the difficult problems...".

The difficult problem would seem to be that there is no simple way to predict load from discharge. Rating curves may give an answer, but the answer is based on an inadequate model in terms of both theoretical considerations and of the complex nature of the relationship between discharge and sediment concentration.

V. VARIATIONS IN SEDIMENT YIELD BEHAVIOUR

A. *Temporal Variations*

An important aspect of the continuous monitoring method of determining sediment yield is that it gives time-based observations of discharge and suspended sediment concentration. These time series can be used to determine the behaviour between the two variables during storm events. Because sediment load is simply the multiplication of discharge and concentration (Eqn. 1), temporal variations in discharge and concentration do affect sediment load.

Hysteresis diagrams are a particularly good technique for determining the temporal behaviour between suspended sediment concentration and discharge or suspended sediment response. Though they contain the same information as simultaneous plots of the hydrograph and sedigraph, they have a number of advantages. First, they give some indication of the distribution, or scatter, between the two variables which should deter most researchers from postulating a simple linear functional relationship. Second, they distinctly show the temporal variability between the two variables which should further negate any desires from linear modelling.

This hysteresis diagram approach has been used to analyse the storm event behaviour between discharge and suspended sediment concentration in five small catchments near Eden, N.S.W. (Olive and Rieger, 1985; Rieger and Olive, 1986). The data used covered a two year period and included 39 storm events. The broad patterns of suspended sediment response types (Fig. 2) indicate the complex variations in the relationship between discharge and concentration.

Sediment leads and lags occurred in 50% of the storms giving rise to relatively small loads due to the different timings of peaks in concentration and discharge. For multiple rising storm event hydrographs, there is evidence of sediment depletion after the initial peak, again resulting in lower sediment loads. Finally, the dominance of the unrecognisable pattern, which occurred in 40% of the storms, adds even more complexity to the nature of sediment loads.

There are also important variations in sediment load at time scales larger than the storm event level. The high

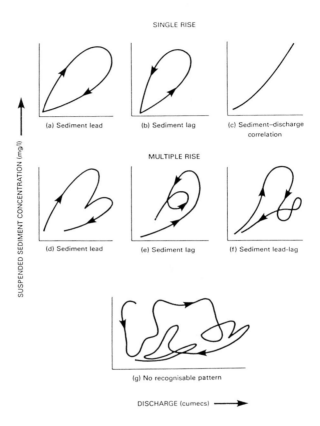

FIGURE 2. *Suspended sediment storm responses (based on Olive and Rieger, 1986).*

variability of Australian precipitation can result in extreme storm events producing dominant sediment loads. Geary (1981) found that 95% of annual sediment load for Deep Creek, N.S.W. could be attributed to storms which occurred during a four week period. For a small experimental catchment over a two year period, Olive and Rieger (1985) estimated that 75% of total load was transported during a single storm event and 92% in two storm events. During a twelve year period, Kata (1978) states that the annual sediment yields of the Ord River ranged from 36.8 to 2104 t $km^{-2}yr^{-1}$.

These long term variations in sediment load point to the problem of the representative nature of sediment yield and denudation estimates. Because yield, itself, is a continuous time series, the inherent variation in the series makes it statistically impossible to calculate a truly representative mean measure. The problem is similar to that faced by meteorologists in attempting to determine a value for mean temperature. The meteorologists arrived at a rather arbitrary solution and simply defined a fixed time period on which mean temperature was to be based. Perhaps a similar approach should be adapted for sediment yields, allowing more realistic comparisons between catchments.

B. Spatial Variations

Within a catchment, there are spatial factors which result in variations in sediment yield. These factors are related to the nature of the sediment delivery process as indicated by Moore (1985). Basically the delivery process involves the interaction between the delivery mechanism and sediment source. The simplest delivery mechanism is overland flow. Sediment delivery only occurs in areas of the catchment where source areas of overland flows overlap sediment source areas. Dunne (1978) has shown that the size and location of discharge source areas change throughout a storm event, depending mainly on the temporal variations of rainfall intensity. Thus, at one time, discharge and sediment source areas will coincide while at another time they will not.

The model is certainly simplistic and ignores such things as pipe flow as a delivery mechanism. However it does point to the important spatial variations in sediment delivery processes which may affect sediment yield within a catchment. It also states that not all areas of a catchment contribute to sediment delivery, a fact which is missed by many researchers.

At a larger spatial scale, there is also spatial variation in sediment yield. The general inverse relationship between sediment yield and catchment size has already been mentioned. The likely reasons for this relationship include the steeper slopes of smaller catchments and storms fully covering small catchments.

A more subtle form of spatial varition has been found between adjacent catchments having relatively the same size, relief, vegetation and geology (Olive and Rieger, 1985). For the same storm event, the adjacent catchments had different sediment responses, with one having a sediment lead while the other had a sediment lag. Such variation should be taken into consideration with experiments using the paired catchment approach. The concept of a representative catchment was originally developed for discharge studies and cannot be extrapolated to sediment yield because of the differences in generation processes.

VI. CONCLUSIONS

Research on sediment yields has been hampered by the continual use of the simple rating curve to estimate channel sediment load. Besides giving inaccurate estimates of load the rating curve has had an insidious effect on its practitioners. They begin to believe that the river catchment is behaving in the same simple fashion as the model they are using to estimate load. Such a belief has clouded the true temporal and spatial variations which occur in the discharge-concentration relationship and in the sediment delivery processes.

Future research on sediment yield should be based on a continuous sampling design which is the only realistic way to estimate sediment load. Turbidity meters, in conjunction with modern data logging equipment, are the simplest method of continuous monitoring. However, turbidity meters are certainly not a "magic bullet" and care must be taken in their application or results could be as misleading as those given by rating curves. Research needs to be carried out on the effects on the turbidity-sediment concentration relationship by such factors as particle size and water colouration.

With more accurate estimates of channel sediment load, all aspects of sediment yield research should be improved compared to its present, sorry state where load data are similar to that generated by random numbers. These accurate sediment load estimates should reveal the inherent spatial and temporal nature underlying sediment transport within stream channels. This, in turn, will give some indication of the location of sediment source areas within the catchment and allow geomorphologists to come to grips with sediment delivery processes, an important research area which has been mostly ignored.

ACKNOWLEDGEMENTS

The authors would like to thank A.R.G.S. for help in funding, the Forestry Commission of N.S.W. for help related to the Eden research area and Dave Gillieson and Roger McLean for their thoughts on denudation rates.

REFERENCES

Abawi, G.Y. and Stokes, R.A. (1982) Wights catchment sediment study 1977-1981. *Water Resources Technical Report No. 100, Public Works Dept.,* West. Aust.
Belperio, A.P. (1979) The combined use of wash load and bed material load rating curves for the calculation of total load: an example from the Burdekin River Australia. *Catena* 6, 317-329.
Chorley, R.J., Schumm, S.A. and Sugden, D.E. (1984) *Geomorphology,* Methuen, London.
Ciesiolka, C.A.A. and Freebairn, D.M. (1982) The influence of scale on runoff erosion. *Inst. Engrs. Aust. Publ.,* 82/8, 203-206.
Douglas, I. (1973) Rates of denudation in selected small catchments in Eastern Australia. Univ. Hull Occasional Papers in Geography, 21.
Dunne, T. (1978) Field studies of hillslope processes. In M. Kirkby (ed.) *Hillslope Hydrology,* Wiley, New York, 227-294.
Ferguson, R.I. (1986) River loads underestimated by rating curves. *Water Resour. Res.,* 22, 74-76.
Field, J.B. (1985) Erosion in a small catchment in New England, NSW, Australia. In R.J. Loughran (ed.) *Drainage Basin Erosion and Sedimentation,* Vol. 2, Univ. Newcastle, NSW, 43-58.
Geary, P.M. (1981) Sediment and solutes in a representative basin. *Aust. Representative Basins Program Report Series No. 3*
Grenney, W.J. and Heyse, E., (1985) Suspended sediment-river flow analysis. *J. Envir. Engr.,* 3, 790-803.
Holeman, J.M. (1968) The sediment yield of major rivers of the world, *Water Resour. Res,* 1, 737-747.
Kata, P. (1978) Ord river sediment study. *Public Works Dept. West. Aust.*
Klemes, V. (1986) Dilettantism in Hydrology: transition or destiny. *Water Resour. Res.,* 22, 177-188.

Leitch, C.J. (1982) Sediment levels in two tributaries of the East Kiewa River pior to logging alpine ash in one of the catchments. *Hydrology, Inst. Engrs. Aust. Publ.,* 82/6, 72-78.

Loughran, R.J. (1969) Fluvial erosion in five small catchments near Armidale, NSW, *Univ. New England, Research Papers in Physical Geography,* No. 1.

Loughran, R.J. (1976) The calculation of suspended sediment transport from concentration v discharge curves - Chandler River, NSW. *Catena* 3, 45-61.

Loughran, R.J. (1977) Sediment transport from a rural catchment in NSW. *J. Hydrol.,* 34, 357-377.

Loughran, R.J. (1984) Studies of suspended sediment transport in Australian drainage basins - a review. In R.J. Loughran (ed.), *Drainage Basin Erosion and Sedimentation,* Univ. Newcastle, NSW, 139-146.

McMahon, T.A. (1982) World hydrology: does Australia fit? *Hydrology and Water Resources Symposium,* Melbourne, Inst. Engrs. Aust., 1-8.

Melton, M.A. (1958) Correlation structure of morphomentric properties of drainage systems and their controlling agents. *J. Geol.,* 66, 442-460.

Moore, R.J. (1985) A dynamic model of basin sediment yield. *Water Resour. Res.,* 20, 89-103.

Olive, L.J. (1973) Sediment yields and stream catchment variations in southeast Tasmania. Unpublished MSc thesis, Univ. Tasmania.

Olive. L.J., Rieger, W.A. and Burgess, J.S. (1980) Estimation of sediment yields in small catchments: a geomorphic guessing game? *Proc. 16th Conf. Inst. Aust. Geogr.* Newcastle, 279-288.

Olive, L.J. and Walker, P.H. (1982) Processes in overland flow-erosion and production of suspended material. In E.M. O'Loughlin and P. Cullen (eds.), *Prediction in Water Quality,* Aust. Academy Science, Canberra, 87-121.

Olive, L.J. and Rieger, W.A. (1985) Variation in suspended sediment concentration during storms in five small catchments in southeast New South Wales. *Aust. Geog. Studies,* 23, 38-51.

Olive, L.J. and Rieger, W.A. (1986) Low Australian sediment yields - a question of inefficient sediment delivery. *IAHS Publ.* 159, 355-366.

Rieger, W.A. and Olive, L.J. (1986) Sediment response during storm events in small forested watersheds. In A.H. El-Shaarawi and R.E. Kwaitkowski (eds.) *Statistical Aspects of Water Quality Monitoring, Developments in Water Science,* 27, 490-398.

Schumm, S.A. (1963) The disparity between present rates of denudation and orogeny. *U.S.G.S Prof. Paper 454-H.*

Strakhov, N.M. (1967) *Principles of Lithogenesis*, Vol. 1, Oliver and Boyd, Edinburgh.

Taylor, G. (1976) The Barwon River, New South Wales - a study of basinfill by a low gradient stream in a semi-arid environment. Unpublished PhD thesis, Aust. National Univ., Canberra.

Walker, P.H. and Hutka, J. (1979) Size characteristics of soils and sediments with special reference to clay fractions. *Aust. J. Soil Res.*, 17, 383-404.

Walling, D.E. (1977) Limitations of the rating curve technique for estimating suspended loads with particular reference to British rivers. *IAHS Publ.*, 122, 34-48.

Walling, D.E. (1978) Reliability considerations in the evaluation and analysis of river loads. *Zeit. f. Geomorph.*, 29, 29-42.

Walling, D.E. (1984) Sediment delivery from drainage basins. In R.J. Loughran (ed.) *Drainage Basin Erosion and Sedimentation*, Newcastle, NSW, 71-80.

Walling, D.E. and Webb, B.W. (1983) Patterns of sediment yield. In K.J. Gregory (ed.) *Background to Paleo-hydrology*, Wiley, New York, 69-100.

5
Determination of Erosion and Accretion Rates using Caesium-137

Robert J. Loughran

Department of Geography
The University of Newcastle, NSW

Bryan L. Campbell

Radioisotope Research Section
Australian Nuclear Science & Technology Organisation
Lucas Heights, NSW

Gregory L. Elliott

Soil Conservation Service of NSW
Research Centre
Gunnedah, NSW

I.   INTRODUCTION

The environmental radioactive tracer, caesium-137, is being used increasingly to provide the "closer tie...between phenomena of erosion from source areas, storage and transport in channel systems", which was called for by Wolman (1977, p.54). Pioneering work on ^{137}Cs in soil erosion and sedimentation studies was begun in the USA by McHenry and Ritchie nearly 20 years ago (Ritchie and McHenry, 1975; McHenry, 1985). Their studies, which demonstrated the utility of the technique, prompted researchers in other countries to investigate its application in a variety of environments. In Australia, the first studies were conducted during the 1970s (Campbell and Ross, 1980; McCallan *et al*,

FLUVIAL GEOMORPHOLOGY OF AUSTRALIA
ISBN   0 12 7356606

1980), and during the 1980s there have been several contributions on the use of ^{137}Cs as a tracer of sediment movement in drainage basins.

Any property or characteristic which makes it possible to identify or to follow the dynamic behaviour of a substance may be considered a tracer. This may take the form of a chemical identity, ionic conductivity, fluorescence, radioactivity or nuclear differences (such as the relative abundance of a particular isotopic mass, ^{16}O/^{18}O, for example). Tracers are quantitative if they meet the following conditions:

(1) do not undergo any transformation when subjected to dynamic action, and
(2) behave dynamically in the same way as the host material when subjected to the same dynamic action.

Caesium (atomic number 55, atomic weight 132.91) is an alkali metal (Group I, period 6). The radioactive isotope caesium-^{137} is a product of a fission reaction (half-life 30.2 years). Atmospheric thermonuclear weapons tests have released significant quantities of ^{137}Cs into the stratosphere, and the isotope has been redistributed over the earth as fallout. On reaching the earth's surface, ^{137}Cs becomes rapidly and firmly adsorbed onto sedimentary particles and further translocation by chemical means is extremely limited (e.g. Davis, 1963; Tamura, 1964; Essington *et al*, 1981; Cerling and Turner, 1982). It therefore becomes an effective tracer of sediment movement.

Caesium-137 levels in soils and sediments may be measured by placing an oven-dried sample in a Marinelli beaker on a hyperpure germanium detector. The net count rate under the ^{137}Cs photopeak (Eγ 0.662 MeV) is a measure of the ^{137}Cs concentration in the sample. Caesium-137 concentration may be expressed per gram of total sample (mBq g^{-1}). However, since ^{137}Cs is preferentially adsorbed onto the finer fraction of soils, concentrations are usually given per unit mass of sample < 2.4 mm or per unit silt + clay (Campbell *et al*, 1982). The ^{137}Cs content at a given site is calculated by dividing the total ^{137}Cs count by the ground surface area of the sampling core or device (in mBq cm^{-2}).

Fallout of ^{137}Cs has been neither spatially nor temporally uniform. Generally ^{137}Cs is found in greater quantities in the northern hemisphere (Davis, 1963). Within Australia, contrasting ^{137}Cs amounts have been noted between south-west Western Australia (50-60 mBq cm^{-2}) and the Hunter Valley, NSW (100 mBq cm^{-2}) (Loughran et al, 1987; Campbell *et al*, 1986b). The temporal pattern in Australia has the following principal features (Campbell, 1983; Campbell *et al*, 1982; Longmore *et al*, 1983):

(1) the first appearance of significant amounts of ^{137}Cs in 1955-56 and rapidly increasing in concentration,
(2) a marked decrease in the rate of deposition from 1959 until 1962,
(3) maximum fallout in 1963-64, and
(4) perturbations due to Chinese and French atmospheric nuclear tests continuing until the early 1980s.

The ^{137}Cs technique for sediment studies depends on:

(1) the fallout of ^{137}Cs from atmospheric nuclear weapons testing, and the lack of other natural sources for this isotope
(2) the strong adsorption of ^{137}Cs on the soil fines
(3) the transport of ^{137}Cs on soils and sediment from eroded sites
(4) the presence of and ability to identify stable, uneroded "input" sites, against which ^{137}Cs losses from eroded sites may be compared
(5) the deposition of labelled material on flood-plains, alluvial fans and slopes, and in reservoirs, lakes and estuaries, with the formation of characteristic ^{137}Cs profiles, and
(6) the ability to detect relatively minute amounts of ^{137}Cs in soils and sediments.

A. *A Model of Caesium-137 Distribution in a Drainage Basin*

Theoretically, the distribution of ^{137}Cs in a drainage basin reflects successive atmospheric inputs and the re-distribution of soil materials by erosion and sedimentation (Fig. 1) (Campbell *et al*, 1982).
Sites experiencing little or no erosion will have ^{137}Cs concentrated in their surface soil layers, usually the upper 5 cm. The total ^{137}Cs amount will represent the total input to the basin, less the loss due to radioactive decay.
On uncultivated hilltops and slopes, sites which have been eroded can be expected to have less ^{137}Cs than nearby input sites. Furthermore, the ^{137}Cs depth-profile may be truncated, especially if the upper layers have been removed progressively thereby preventing build-up of the isotope over time.

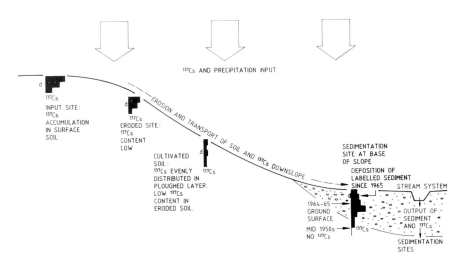

FIGURE 1.   Model of caesium-137 redistribution on drainage basin sediments.

Cultivated sites should have ^{137}Cs relatively evenly distributed within the plough layer. If the soil has been eroded, ^{137}Cs amounts should be lower than input values.

At active sedimentation sites, ^{137}Cs may vary with depth and in amount. Sediments containing no ^{137}Cs can be assumed not to have been exposed to fallout and, in a sedimentary depth profile, may be dated as pre-1954. Peaks in depth concentrations can be correlated with ground surfaces exposed to maximum fallout in 1963-64; to these could be added labelled material from other sites then undergoing erosion. Total ^{137}Cs at a sedimentation site can be expected to be greater than input, having received ^{137}Cs from both fallout and sediment deposition. However, cut-and-fill and particle size effects due to sorting may occasionally reduce ^{137}Cs levels.

II. DETERMINATION OF EROSION FROM CAESIUM-137 MEASUREMENTS

Relative levels of ^{137}Cs, and therefore soil loss, must be assessed against an input value for the drainage basin. The choice of sites for input sampling is critical. Open sites on hilltops, undisturbed by animal or human activity, are preferred. Forested hilltops, although probably stable, may have been labelled unevenly due to the variable effects of canopy cover, wind funelling, leaf drip, stemflow, etc.

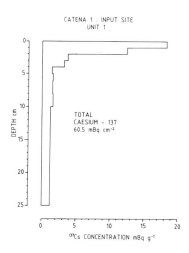

FIGURE 2. Caesium-137 distribution at an input site, near Narrogin, W.A.

Results from Maluna Creek basin in the Hunter Valley, NSW, indicate some variation in ^{137}Cs values at potential input sites in forest (Loughran *et al*, 1986a). Open sites, even though grazed by cattle, have given more consistent results (Campbell *et al*, 1986b). Normally, at least three sites are sampled in close proximity, one by depth-increments; the remainder by coring (Campbell *et al*, 1982; 1986b). Mean ^{137}Cs areal activity of these sites is taken as the input value (McCallan *et al*, 1980).

A typical input depth-profile is shown in Figure 2 for an open woodland site 40 km east of Narrogin in the Wheat Belt of Western Australia (Loughran et al, 1987). The greater part of the ^{137}Cs was in the top two centimetres of the soil, with concentrations below 5 cm approaching background activity. The apparently uniform concentration of ^{137}Cs between 10 and 25 cm was due to bulk sampling by core. The total ^{137}Cs content at this site was 60.5 mBq cm^{-2}, compared with 59.5 and 46.1 mBq cm^{-2} at neighbouring sites, bulk sampled by core. The lowest value may be due to disturbance or uneven labelling. A realistic input value in this situation would be 60 mBq cm^{-2}.

As more ^{137}Cs measurements are made, the spatial variation (or lack of it) of input values for Australia is becoming apparent. For example, Longmore *et al* (1983) reported 87 mBq cm^{-2} for the south-east Darling Downs,

Campbell et al (1986b) 100 mBq cm^{-2} for the eastern Hunter Valley, 89 mBq cm^{-2} for Merriwa in the upper Hunter, 89-101 mBq cm^{-2} for Manilla, NSW, and 90 mBq cm^{-2} for the central tablelands of NSW near Goulburn. In northwestern Victoria near Lake Tyrrell, Longmore et al (1986b) reported an input value of ^{137}Cs of 77 mBq cm^{-2}. Longmore et al (1983) estimated fallout at Brisbane (corrected for decay to 1978) to be 86 mBq cm^{-2}. This indicates that ^{137}Cs fallout may not be closely related to rainfall quantity in central eastern Australia (Longmore et al, 1983).

Uncultivated sites which have experienced sheet and rainsplash erosion exhibit truncated ^{137}Cs profiles and lower amounts of the isotope per unit area. A site under open woodland, near Narrogin, WA, shows such characteristics (Fig. 3). Compared with the input site (Fig. 2), there has been a loss of nearly 30% ^{137}Cs (Loughran et al., 1987). Independent measurements of soil movement here by Pilgrim (1981) over a 6.2 year period showed soil losses of 69 kg ha^{-1} y^{-1}.

Cultivated soils have ^{137}Cs redistributed within the plough layer. Figure 4 illustrates this effect for a brown podzolic soil in the Maluna Creek catchment at Pokolbin. Furthermore, the lower amount of ^{137}Cs in the cultivated soil indicated more severe soil erosion, which was confirmed by a survey of four soil types under vineyards and grazing land in that drainage basin (Loughran *et al*, 1982).

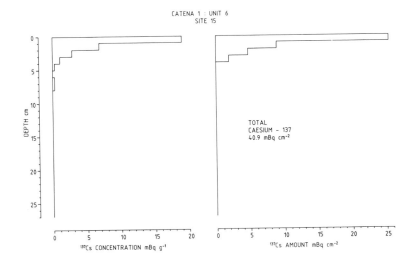

FIGURE 3.   Caesium-137 distribution at a slope site, Narrogin, W.A.

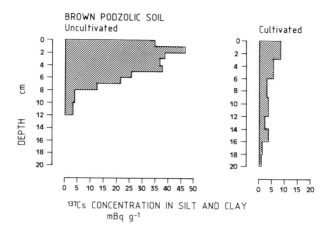

FIGURE 4. Caesium-137 distribution on brown podzolic soil, Maluna Creek catchment, NSW.

The relationship between erosion and ^{137}Cs loss was used by Longmore *et al* (1983) to map zones of soil loss and gain in a 33 ha cultivated paddock on the Darling Downs, Queensland. Soil samples were collected on a 100 m parallel grid (three samples per site), ^{137}Cs values were plotted on a map and lines of equal ^{137}Cs content (isocaes) drawn. The emergent pattern shows distinct zones of ^{137}Cs depletion (lower than input=soil erosion) and accumulation (higher than input=soil deposition) (Longmore *et al*, 1983). Semi-quantitative grades of erosion severity were based on the degree of ^{137}Cs loss. Similarly, net soil deposition was graded according to the quantity of ^{137}Cs present above the input value.

Quantification of soil loss by ^{137}Cs measurements became possible with the use of erosion data from plots of the Soil Conservation Service of NSW and the University of Newcastle. Caesium-137 was measured in soil cores taken from within 100 m² plots or adjacent to 2 m² plots. Comparable soil types on uneroded sites were used to estimate ^{137}Cs input and the degree of ^{137}Cs loss from the plots (as a percentage of input) (Campbell *et al*, 1986a). The relationship between net soil loss and ^{137}Cs loss is

$$Y = 0.419 \; X^{0.646} \quad (r = 0.89, \; n = 30) \tag{1}$$

where Y is the diminution of ^{137}Cs content and X is the average soil loss in kg ha^{-1} y^{-1}.

Mapping of net soil loss was first attempted in Maluna Creek catchment. Two vineyard blocks (each 0.3 ha in area) were sampled on a 10 x 10 m grid and maps of isocaes drawn (Campbell *et al*, 1986b). Net soil loss (kg ha^{-1} y^{-1}) was estimated from the two-part relationship shown in Figure 5, and the isocaes were translated into lines of equal net soil loss, or isoerosols. Figure 6 shows the results from one of the vineyards, where the net soil loss was estimated as 5.9 t ha^{-1} y^{-1} (Loughran *et al*, 1986c).

Established techniques for estimating soil erodibility have been tested against ^{137}Cs content (Elliott *et al*, 1984). While many of the indices were poorly correlated, the soil aggregate stability index of Collis-George and Figueroa showed a good relationship (Fig. 7a), indicating that soils with more stable aggregates had higher ^{137}Cs concentrations in the silt + clay fraction. Caesium-137 concentration was also related to soil organic matter (Fig. 7b, r = 0.53). This suggested that the more eroded soils had lower ^{137}Cs concentrations and less organic matter, or that higher organic content was an indicator of increased aggregate stability, thus making the soil less susceptible to erosion.

Caesium-137 concentrations on stream-borne sediments can be used as an indicator of drainage basin sediment

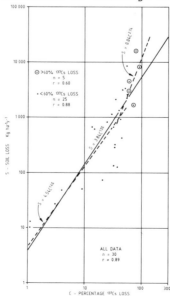

FIGURE 5. Net soil loss v. caesium-137 loss, eastern NSW.

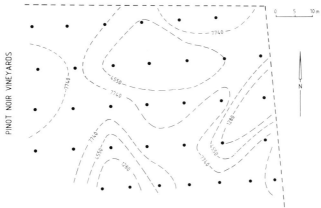

FIGURE 6. Maluna vineyard with superimposed isocaes mBq cm^{-2} (top), and map of net soil loss kg ha^{-1} y^{-1} (below).

source. This is the so-called "fingerprinting" technique (Walling and Kane, 1984). The method relies on significant variations in average ^{137}Cs concentration occurring according to sediment sources within the basin. For example, in Maluna Creek catchment, stream-borne sediment delivered to the basin outlet had a ^{137}Cs concentration of 4.6 ± 0.6 mBq g^{-1}. Concentrations of ^{137}Cs on cultivated and uncultivated soils in the basin were, on average, 2.7 and 30.2 mBq g^{-1}, respectively (Loughran *et al*, 1986a). To

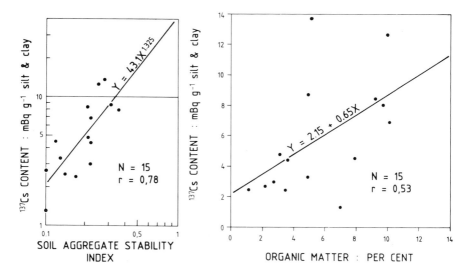

FIGURE 7. *Caesium-137 content in relation to (a) soil aggregate stability index, and (b) soil organic matter.*

achieve an output ^{137}Cs concentration of 4.6 mBq g^{-1}, a 93% contribution is required from cultivated land, and 7% from uncultivated (forest and grassland). When particle-size effects were taken into consideration, since the transported sediment was enriched in clay compared with potential sources, the cultivated vineyards were considered to be the sole source of sediment to the stream system (Loughran *et al,* 1986a).

III. DETERMINATION OF ACCRETION RATES FROM CAESIUM-137
      MEASUREMENTS

The deposition of ^{137}Cs directly from fallout and on incoming sediments can lead to the development of distinctive ^{137}Cs profiles (Fig. 1). Use of ^{137}Cs as a dating-tool for the profile depends on the rate of incoming ^{137}Cs and sediments to provide measurable variations in ^{137}Cs with depth. Very low rates of accretion (millimetres per year rather than centimetres per year) mean that any diffusion of ^{137}Cs greatly blurs finer detail (McHenry, 1985).

There are three possible mechanisms by which ^{137}Cs can be diffused up or down a profile following deposition:

(1) chemical diffusion
(2) mixing, and
(3) faunal mixing.   (McHenry, 1985).

It is generally considered that ^{137}Cs is not affected by chemical diffusion in a sediment column (Wise, 1980), although Longmore (1982) and Longmore et al (1986a) believe ^{137}Cs diffusion can take place in highly saline or acid conditions. Mixing due to runoff or wave action in a shallow waterbody may disturb sedimentary profiles, and where animals are present, disturbance by trampling and feeding can occur.

Floodplains, alluvial fans, channels and reservoirs are the chief locations of sediment storage in drainage basin systems. Caesium-137 has been used to determine sedimentation depths and rates in all of these environments. Within Maluna Creek catchment, soil erosion from vineyard sources has promoted the post-1970 development of flood-plain, channel and alluvial fan deposits (Campbell et al, 1982; Loughran and Campbell, 1983). The ^{137}Cs profile for an alluvial fan (Fig. 8a) shows a maximum concentration in the 63-65 cm layer, which probably represents a surface exposed at the time of maximum fallout, plus the deposition of strongly labelled material derived from erosion of adjacent vineyards established in 1970. Concentrations of ^{137}Cs decrease up the profile owing to the deposition of progressively weaker labelled soils (Campbell et al, 1982). A shallower ^{137}Cs profile was evident on the floodplain of Maluna Creek, where ^{137}Cs was concentrated in the upper 13.5 cm (Fig. 8b). It was concluded from ^{137}Cs and sedimento-logical evidence that approximately 8.5 cm of sediment had accumulated since 1965 (Campbell et al, 1982). Sediment which has been deposited in Maluna Creek channel also exhibited a ^{137}Cs profile (Fig. 9). Again, the ^{137}Cs peak concentration was at depth (54-63 cm), with no ^{137}Cs present below 73 cm, which suggested that approximately 60 cm of sediment had been deposited since the time of maximum fallout and the inception of the vineyards (1965-70). Furthermore, the total ^{137}Cs content at the channel site was more than 3.5 times the input value for the catchment, confirming significant sediment accumulation at this site.

Sedimentary deposits behind water storages of various sizes have been dated by the ^{137}Cs technique (Loughran and Campbell, 1983; Campbell, 1983; Day and Campbell, 1986). Figure 10 shows the ^{137}Cs sequence for sediments behind Wyangala dam on the Lachlan River, NSW. Above this point the catchment area is 8320 km^2. Caesium-137 makes its first appearance at 250 mm depth (1955) and peaks at 105 mm

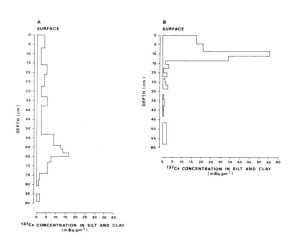

FIGURE 8.   Caesium-137 distribution at sedimentation
sites in Maluna Creek catchment, NSW.   (a) Alluvial fan;
(b) Floodplain.

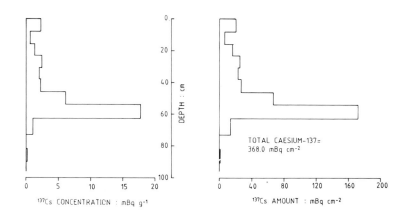

FIGURE 9. Caesium-137 distribution on channel-fill
sediment, Maluna Creek catchment, NSW.

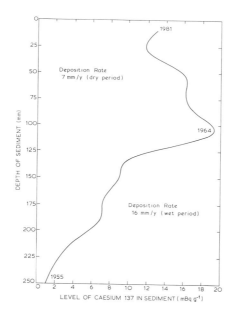

FIGURE 10. Caesium-137 distribution of sediments from Wyangala Dam, Lachlan River, NSW.

(1964). The interval represented by the profile has been subdivided into a "wet period", with a deposition rate of 16 mm y^{-1}, and a "dry period", with a deposition rate of 7 mm y^{-1} (Campbell, 1983, in collaboration with D. Outhet, NSW Department of Water Resources).

The quantity of ^{137}Cs deposited at sedimentation sites partly depends on sorting and particle-size effects. In Maluna Creek basin, the amount of ^{137}Cs in a sand-rich alluvial fan was disproportionately lower than in silt and clay sediments trapped behind a farm dam downstream (Loughran *et al*, 1986b). This factor, as well as possible disturbances by cultivation and animal activity, must always be taken into consideration when interpreting sedimentation sites.

IV. CONCLUSION

Within drainage basins, soil erosion status may be indicated by levels of adsorbed ^{137}Cs within the soil (eg Loughran et al, 1982; Elliott *et al*, 1983; Reece and Campbell, 1984; Day and Campbell, 1986). Further refinements of the relationship between soil loss and ^{137}Cs loss (Fig. 5) for various soil types, geographic locations and for soils experiencing severe erosion will enable more accurate predictions of net soil loss rates to be made. For a long time quantification of sediment sources within catchments has been a difficulty for hydrogeomorphologists interpreting sediment outputs, and the ^{137}Cs techique holds much promise for its resolution.

Storage, another major component of the sediment budget, may also be quantified by ^{137}Cs, provided careful sampling and interpretation are practised. Caesium-137 as a dating-tool is less useful at sites with low rates of sediment accumulation, ostensibly those unaffected by human activity in the catchment (McHenry, 1985). "Correct and careful interpretation of ^{137}Cs data is much more of a problem than actually determining the presence or concentration of [137]Cs" (McHenry, 1985, p.39).

The tracer ^{137}Cs has become a well-established tool in sediment research in several countries. As well as in USA and Australia, it has been used in Canada by De Jong and colleagues as an indicator of soil erosion (e.g. De Jong *et al*, 1982). Walling and co-workers in England have used it as a "fingerprint" on suspended sediment in order to determine drainage basin sediment sources (e.g. Peart and Walling, 1986), and sedimentation rates in Lake Geneva have been measured from ^{137}Cs profiles by Vernet *et al* (1984). The provision of adequate detector facilities will ensure the further development of the technique in Australia, thus helping hydrogeomorphologists achieve the goal of predicting sediment delivery ratios and drainage basin sediment budgets with some certainty.

REFERENCES

Campbell, B.L. (1983) Applications of environmental caesium-137 for the determination of sedimentation rates in reservoirs and lakes and related catchment studies in developing countries. *Int. Atomic Energy Agency*, Vienna, Tech-Doc 291.

Campbell, B.L. and Ross, M. (1980) Determination of sedimentation rate in Stephens Creek reservoir using environmental caesium-137. *Conf. of Inst. Aust. Geogrs. Papers*, Newcastle, 43-51.

Campbell, B.L., Elliott, G.L. and Loughran, R.J. (1986a) Measurement of soil erosion from fallout of Cs-137. *Search*, 17, 148-149.

Campbell, B.L., Loughran, R.J. and Elliott, G.L. (1982) Caesium-137 as an indicator of geomorphic processes in a drainage basin system. *Aust. Geogr. Stud.*, 20, 49-64.

Campbell, B.L., Loughran, R.J., Elliott, G.L. and Shelly, D.J. (1986b) Mapping drainage basin sediment sources using caesium-137. *Drainage Basin Sediment Delivery*, Albuquerque Symposium, IAHS Publ. No. 159, 437-446.

Cerling, T.E. and Turner, R.R. (1982) Formation of freshwater Fe-Mn coatings on gravel and the behaviour of Co-60, Sr-90 and Cs-137 in a small watershed. *Geochim et Cosmochim Acta*, 46, 1333-1343.

Davis, J.J. (1963) Caesium and its relationship to potassium in ecology. In V. Schultz and A.W. Klements (eds.) *Radioecology*, Reinhold, New York, 539-556.

Day, D.G. and Campbell, B.L. (1986) Environmental caesium-137 as an indicator of accelerated erosion and sedimentation in Birchams Creek, NSW. *ANU Centre for Res. Envir. Studies, Working Paper*, 1986/42.

De Jong, E., Villar, H. and Bettany, J.R. (1982) Preliminary investigations on the use of Cs-137 to estimate erosion in Saskatchewan. *Can. J. Soil Sci.*, 62, 673-683.

Elliott, G.L., Campbell, B.L. and Loughran, R.J. (1984) Correlation of erosion and erodibility assessments using caesium-137. *J. Soil Cons. NSW*, 40, 24-29.

Elliott, G.L., Lang, R.D. and Campbell, B.L. (1983) The association of tree species, landform, soils and erosion on Narrabeen sandstone West of Putty, New South Wales. *Aust. J. of Ecology*, 8, 321-331.

Essington, E.H., Fowler, E.B. and Polzer, W.L. (1981) The interactions of low-level, liquid radioactive wastes with soils. 2. Differences in radionuclide distribution among four surface soils. *Soil Sci.*, 132, 13-18.

Longmore, M.E. (1982) The caesium-137 dating technique and associated applications in Australia - a review. In W. Ambrose and P. Duerden, (eds.), *Archaeometry - an Australian Perspective*, ANU Press, Canberra, 310-321.

Longmore, M.E., O'Leary, B.M., Rose, C.W., and Chindica, A.L. (1983) Mapping soil erosion and accumulation with the fallout isotope caesium-137. *Aust. J. Soil Res.*, 21, 373-385.

Longmore, M.E., Torgersen, T., O'Leary, B.M. and Luly, J.G. (1986a) Caesium-137 redistribution in the sediments of the playa, Lake Tyrrell, northwestern Victoria. I. Stratigraphy and caesium-137 mobility in the upper sediments. *Palaeogeogr., Palaeoclimatol., Palaeoecol.*, 54, 181-195.

Longmore, M.E., Luly, J.G. and O'Leary, B.M. (1986b) Caesium-137 redistribution in the sediments of the playa, Lake Tyrrell, northwestern Victoria. II. Patterns of caesium-137 and pollen redistribution. *Palaeogeogr., Palaeoclimatol., Palaeoecol.*, 54, 197-218.

Loughran, R.J. and Campbell, B.L. (1983) The determination of sedimentation depth by caesium-137. *Search*, 14 157-158.

Loughran, R.J., Campbell, B.L. and Elliott, G.L. (1982) The identification and quantification of sediment sources using Cs-137. *Recent Developments in the Explanation and Prediction of Erosion and Sediment Yield*, Exeter Symposium, IAHS, Publ. No. 137, 361-369.

Loughran, R.J., Campbell, B.L. and Elliott, G.L. (1986a) Sediment dynamics in a partially cultivated catchment in New South Wales, Australia. *J. Hydrol.*, 83, 285-297.

Loughran, R.J., Campbell, B.L. and Elliott, G.l. (1986b) The use of the tracer caesium-137 for studying sediment movement in a drainage basin. *Pub. Geol. Soc. Aust. NSW Div.* No. 2, 151-160.

Loughran, R.J., Campbell, B.L. and Elliott, G.L. (1986c) A nuclear technique measures soil erosion. *Nuclear Spectrum*, 2, 2-4.

Loughran, R.J., Campbell, B.L., Pilgrim, A.T. and Conacher, A.J. (1987) Caesium-137 in soils in relation to the nine-unit landsurface model in a semi-arid environment in Western Australia. In A.J. Conacher (ed.) *Readings in Australian Geography*, Proc. 21st I.A.G. Conf., Perth, May 1986, 398-406.

McCallan, M.E., O'Leary, B.M. and Rose, C.W. (1980) Redistribution of caesium-137 by erosion and deposition on an Australian soil. *Aust. J. Soil Res.*, 18, 119-128.

McHenry, J.R. (1985) Quantification of soil erosion and sediment deposition - the future. USA experience and its relevance to world needs. *Conf. and Review Papers, Drainage Basin Erosion and Sedimentation*, Vol. 2, Univ. of Newcastle, 33-41.

Peart, M.R. and Walling, D.E. (1986) Fingerprinting sediment sources: the example of a drainage basin in Devon, UK. *Drainage Basin Sediment Delivery*, Albuquerque Symposium, IAHS Publ. No. 159, 41-55.

Pilgrim, A.T. (1981) Spatial variability of hydrologic response on naturally vegetated hillslopes in a semi-arid environment. Unpublished PhD thesis, Univ. of Oklahoma, Norman.

Reece, P.H. and Campbell, B.L. (1984) The use of Cs-137 for determining soil erosion differences in a disturbed and non-disturbed semi-arid ecosystem. *Proc. 2nd Int. Rangelands Congr.*, Adelaide.

Ritchie, J.C. and McHenry, J.R. (1975) Fallout Cs-137: a tool in conservation research. *J. Soil and Water Cons.*, 30, 283-286.

Tamura, T. (1964) Selective sorption reaction of caesium with mineral soil. *Nuclear Safety*, 5, 262-268.

Vernet, J.P., Dominik, J. and Favarger, P.Y. (1984) Texture and sedimentation rates in Lake Geneva. *Env. Geol.*, 5, 143-149.

Walling, D.E. and Kane, P. (1984) Suspended sediment properties and their geomorphological significance. In T.P. Burt and D.E. Walling (eds.) *Catchment Experiments in Fluvial Geomorphology*, Geo Books, Norwich, 311-334.

Wise, S.M. (1980) Caesium-137 and lead-210: a review of the techniques and some applications in geomorphology. In R.A. Cullingford, D.A Davidson and J. Lewin (eds.), *Timescales in Geomorphology*, Wiley, Chichester, 109-127.

Wolman, M.G. (1977) Changing needs and opportunities in the sediment field. *Water Resour. Res.*, 13, 51-54.

6
Modelling Arid Zone Soil Erosion at the Regional Scale

Geoff Pickup

CSIRO
Division of Wildlife and Rangelands Research
Alice Springs,   NT

I. INTRODUCTION

The most widespread effect of Europeans on the geomorphology of Australia results from the activity of
grazing animals. This effect has been greatest in the arid
zone where 880,000 km^2 of the 3.36 million km^2 used for
grazing is experiencing soil erosion and adverse changes
in vegetation, and of this, 150,000 km^2 is classed as
severely eroded (Woods, 1983). It is therefore surprising
that the problem of arid zone soil erosion has been all
but ignored by Australia's geomorphologists. The result
is that little is known about the basic processes of soil
erosion, how, when and where it occurs, and what its consequences are for soil productivity, plant communities and
catchment runoff.

Most research into soil erosion has concentrated on
point processes. Less work has been done on spatial and
temporal behaviour, the result being that it is proving
difficult to extend process models from the point to the
region. This difficulty may be of little significance
for erosion modelling in croplands where results are
required at the field scale. It is, however, a crucial
problem in arid grazing lands where information is required
at the paddock or regional scale for land management. This
essay discusses how the problem may be overcome and
examines progress in developing models of erosion which
operate at the regional scale. It begins with a discussion

FLUVIAL GEOMORPHOLOGY OF AUSTRALIA
ISBN   0 12 735660 6

of conventional approaches to erosion modelling which are
largely derived from studies of point processes and notes
their limited capacity for extrapolation to large areas.
Sedimentation processes operating at the regional scale
are then examined leading to the formulation of regional
models capable of describing major landscape changes.

II. REQUIREMENTS OF ARID ZONE SOIL EROSION MODELS

In croplands, soil erosion models are required to
produce a rate of soil loss for a specific soil, crop, or
cultivation practice in a small relatively uniform area
from a particular rainfall. In the arid zone, the interest
is not so much in soil loss *per se*; instead a model should
produce information on changes in biological productivity
over a period of several years for a large and highly
diverse area. These differences result from the purpose
for which the information is required, the type of land
use involved, the land management techniques available,
and the spatial scale at which management is carried out.
Consider, firstly the type of information required from
a model. The principal land use is grazing which relies
on the quality and quantity of natural vegetation. Changes
in the vegetation occur for a number of reasons including
the gain or loss of soil and nutrients, moisture supply
from rainfall and runon, the extent of soil degradation
factors such as compaction and sealing through the development of surface crusts, and so on. Information on soil
loss is, in itself, not sufficient to predict changes in
vegetation. At the same time, however, moisture supply,
soil and nutrient supply and soil degradation are closely
related to soil loss or gain. It may therefore be better
to produce information on changes in soil productivity as
reflected by the vegetation in an area from a model rather
than soil loss alone.
A second difference between arid zone and cropland soil
erosion models is the level of precision required from the
output. In arid grazing lands, individual paddocks can
be hundreds or even thousands of square kilometres in size
and the only cheap way of managing soil erosion is to change
the distribution and severity of grazing and trampling.
This may be done by relocating or adding fences, changing
the location of waterpoints or changing animal numbers. The
effect of these changes on animal behaviour and paddock
usage can only be predicted approximately so there would

be little point in modelling soil loss or gain precisely at every point in the paddock even if such precision was technically possible.

Another difference between arid zone and cropland soil erosion models is the scale at which they need to operate. Croplands are usually modelled in small units of 1 hectare or less on the basis of individual storms, sequences of storms, or design rainfalls of short duration. This makes it possible to treat each area as being spatially uniform and unchanging through time. In the arid zone, a model needs to handle large, highly diverse areas and to produce information on short or medium term trends rather than individual events. It must, however, be capable of showing what these trends are in different parts of the system which cannot, therefore, be treated as spatially uniform.

III. PROBLEMS IN ADAPTING FIELD SCALE MODELS TO LARGE AREAS

Mathematical models of soil erosion occupy points along a continuum between two basic types: the lumped parameter black box model and the distributed process model. Black box models consist of one or more equations which transform values of a number of input variables into some output quantity. They are simple and are formulated without reference to the internal workings of the physical system they represent. The few parameters they contain average system behaviour over space, time or both, hence the term "lumped". Such models have to be derived empirically from measurements of input and output quantities from the system under study or from a prototype. Most black box models relate sediment yield at a catchment outlet point to rainfall or runoff averaged over the whole area. Examples of such models are regression equations, Box-Jenkins (1976) time series models, and spectral transfer function models (Rieger *et al*, 1982). If enough data are available, black box sediment yield models may sometimes be regionalised, the classic case being the Universal Soil Loss Equation (Wischmeier and Smith, 1978) whose basis is a set of regression equations derived from many years of erosion plot experiments in the US. There is little chance of accumulating enough data for such a relationship in Australia from plots. It may, however, be possible to derive some sort of regional relationship from rainfall simulators (Edwards and Charman, 1980).

Lumped parameter black box sediment yield models can, theoretically, be applied at any spatial or temporal scale provided enough data are available for fitting. In reality, this is not so and their usefulness is restricted by lack of data, increasingly complex behaviour as the system becomes larger, and difficulties in deriving meaningful input parameters. Consider firstly, the problem of data. Collecting information on runoff and sediment yield is expensive, labour-intensive and time-consuming, especially in the arid zone where surface runoff occurs infrequently. These restrictions make it impossible to develop black box models empirically within a reasonable time at acceptable cost. The second issue is complexity. Large catchments contain many sources of sediment, and a variety of paths and sinks which sediment must pass through before reaching the outlet where it is measured. Their behaviour also varies with time so it is unlikely to be represented by a few very simple mathematical relationships. The third problem restricting the application of black box models is the difficulty of obtaining representative input data. As the area becomes larger, spatially lumped or averaged input data are increasingly unrepresentative of what is occurring. When this happens with several different input variables which do not covary spatially, the effect is com-pounded and the input data have even less predictive value.

Distributed process models provide an alternative to black box structures and may eventually replace them. These models consist of a set of analytically-based relationships for the determination of overland flow from rainfall coupled with an erosion-deposition process model (see, for example, Foster, 1982; Rose *et al*, 1983a). The erosion process model usually consists of a rainsplash or rainfall detachment model, a sediment transport equation, and an erosion-deposition function such as the equation for con-servation of mass. The erosion-deposition model reflects the current knowledge of the processes involved and is partly analytical and partly empirical. The basic equations are usually solved for a planar land element with uniform soil and hydraulic characteristics although solutions for more complex surfaces exist. Complex topography is, however, normally handled by cascading a series of sub-models and routing the water and sediment flux through them. Ideally, a distributed process model should be capable of producing continuous data on erosion rates through time from measurable physical parameters. In practice, such models contain parameters which must be calibrated by comparing observed behaviour with model output. They are also subject to timing errors such that, even after calibration, observed and modelled behaviour are frequently not in phase (see, for example, Borah *et al*, 1982; Rose *et al*, 1983b).

Distributed process models should, in theory, be far more capable of extrapolation to larger spatial scales as long as the appropriate input data can be obtained. In practice, this is not the case because some elements of erosion and deposition are still treated as point processes which have to be extrapolated to larger areas by spatial averaging. This type of extrapolation cannot be taken very far because, as the area modelled becomes larger, its diversity increases. It is then necessary to represent the diversity with cascaded submodels and to represent the interactions between them with water and sediment routing procedures. This approach allows the behaviour of the system to be built up as the summation of the behaviours of its constituent parts. It has the disadvantage that submodels and routing procedures may contain systematic errors which compound with each step through the model submerging the regional trends which the model is trying to represent. Further difficulties are imposed when the model has to be calibrated. Almost invariably, calibrated parameters are specific to the time and distance steps used in a model (see, for example, Higgins, 1976) and cannot be applied at other spatial or temporal scales. Distributed process models are therefore just as constrained in their use by data availability as black box models.

IV. REGIONAL PROCESS MODELS

An alternative approach to the behaviour of large systems is to develop regional process models which generalise local change but reproduce regional differences and trends. Such models rely on the assumption that, while a system's localised behaviour may be complex, the rules governing the way its component parts interact are simple. However, when those interactions are summed, a behaviour emerges which is a property of the system as a whole. The regional model is therefore intended to build the complex behaviour of whole systems by aggregating a set of simple relationships describing local properties and behaviour.

There are a number of ways to build a regional process model. The approach used in this essay owes much to time series analysis and to methods derived from the analysis and generation of textures in image processing. The

regional process model reproduces system behaviour in
detail on a grid cell basis and is derived by separating
that behaviour into a number of component parts. These
are:

(1) a local response model describing the interaction
between adjacent cells and capable of reproducing that
interaction as the system becomes more eroded;
(2) a trend model to describe systematic regional
variations not handled by the local response model.

It may also be necessary to have a procedure for subdividing
the system into areas in which different local response
models may be applied. Some of the procedures described
have been implemented as part of the micro-BRIAN image
processing package (CSIRO/MPA, 1986; Pickup and Chewings,
1986a). Others are still under development so this essay
is in effect only an interim report.

Every model must, of course, reproduce the behaviour
of a state variable. The model described here was
developed for use in relatively flat arid and semi-arid
lands and uses the erosion-deposition index developed by
Pickup and Nelson (1984) from LANDSAT data. Low values
of the index indicate soil loss, intermediate values are
characteristic of stable areas, and high values are
associated with deposition. The index has been tested
widely in central Australia but would not necessarily apply
in areas other than flat arid lands. There is, however,
no reason why the modelling strategy could not be applied
to other data such as elevations or amounts of soil gain
and loss determined by stratigraphic methods.

A. *Regional Scale Processes*

Model structure and behaviour should, so far as possible,
resemble the physical system they are supposed to represent.
Model development therefore begins with a discussion of
how sedimentation processes operate in aggregate over large
areas.

At the regional scale, sediment is not detached,
entrained and transported smoothly down a system or in time
as most models suggest. Instead, eroded sediment is
supplied discontinuously from a restricted number of
locations and transported intermittently at rates which
are more related to supply than to transport capacity.
Transported sediment may also spend long periods in storage

awaiting the occurrence of an event large enough to remobilise it. In spite of this complexity there is still some order in regional scale sediment transport. This order occurs because sediment seems to move in a series of scour-transport-fill (STF) sequences which appear in the landscape as alternating erosion and deposition (Fig. 1). STF sequences have been identified in a wide range of environments including experimental flumes (Pickup, 1974), gullies and desert arroyos (Leopold *et al*, 1964; Blong, 1966; Schumm and Hadley, 1957), small streams dominated by snowmelt (Meade, 1985), gravel bed rivers (Church and Jones, 1982), and desert alluvial fans and footslopes (Pickup, 1985). They may also produce the chain of ponds features described by Eyles (1977). The phenomenon appears to be fairly general but its mechanics are not well understood. It is, however, essential to an understanding of processes at the regional scale and forms the basis for the local response model.

STF sequences arise from an imbalance between transport capacity and sediment supply with respect to distance and time. This may result from extrinsic factors such as downstream variations in slope, soil type, or geological structure whereby sediment is delivered from point sources into a system whose transport capacity declines downstream. It is, however, also intrinsic to the process of sediment transport from a point source in unsteady flow. Water travels faster than all but the finest fraction of the sediment it transports so a flood wave tends to overtake a pulse of sediment from a point source along its path, resulting in deposition. Many of these deposits may be moved farther downstream in the next flood. However, if they are large enough, they may result in backup slowing the flow down and causing new deposition at their upstream end. The STF sequences then becomes fixed or self-enhancing, a process which is aided by packing of the deposits, washing of clay particles into interstices, and the growth of vegetation, all of which increase resistance to subsequent erosion.

These sequences can exist at a wide variety of amplitudes and spatial scales depending on the magnitude of the events which generate them. They are particularly well-developed in flat arid and semi-arid environments where overland flow is an important process of sediment transport. Under these circumstances, the whole landscape can be made up of such sequences separated by a few stable areas as Pickup (1985) has shown with the development of erosion cell mosaics.

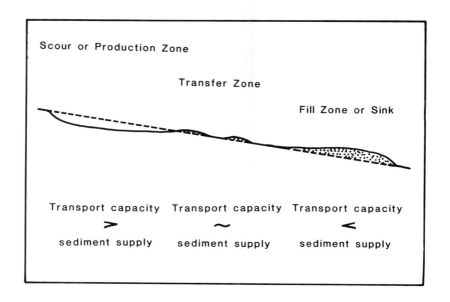

Scour or Production Zone

Transfer Zone

Fill Zone or Sink

Transport capacity   Transport capacity   Transport capacity

>   ~   <

sediment supply   sediment supply   sediment supply

FIGURE 1.   Schematic diagram of a scour-transport-fill (STF) sequence.

The spatial patterns produced by STF sequences vary enormously.   In arid lands, at very low gradients, the size, shape and location of these sequences may be apparently random, producing an erosion cell mosaic (Fig. 2).   Where slopes are greater and a drainage net has developed, it is common to find sediment production zones in the upper and middle interfluves, while the channels are choked with fill (Fig. 2).   The lower interfluves act as transfer zones and are covered by a thin, discontinuous mantle of sediment in temporary storage.   On large arid zone rivers, the sequences appear as sections of bare eroded floodplain separated by massive splays of material which have spilled out of the channel and been deposited.   The features may be up to several kilometres in length.

The permanency of STF sequences may also vary.   In ephemeral rivers, for example, small sequences may be very short-lived, only surviving between floods.   The larger ones which result from rare events or extrinsic factors are maintained for very long periods.   In arid environments, where depositional areas are highly favoured sites for plant growth, once such a sequence has developed and the

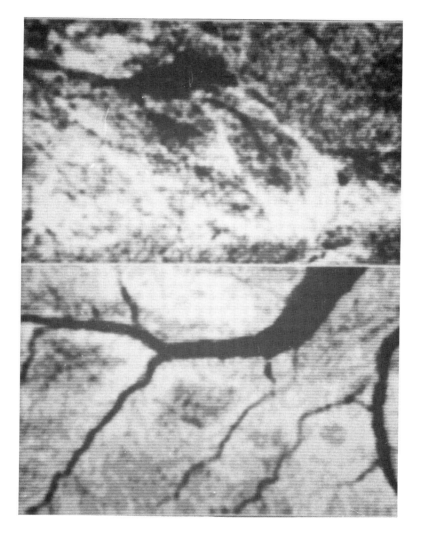

FIGURE 2. Grey scale images showing patterns of erosion and deposition created by multiple STF squences. The upper image illustrates the erosion cell mosaic structure which develops in flat arid and semi-arid alluvial fan and footslope country. The lower image indicates the pattern common in steeper country where tributary rather than distributary drainage occurs. Light tones indicate erosion, dark tones indicate deposition while intermediate tones show areas of relative stability. Both images have been smoothed to enhance the major structures.

depositional area has been colonised by plants, it can survive
for a very long period.

Persistent STF features, once developed, can strongly
influence future landscape behaviour through the process
of buffering. Buffering is at its most extreme in flat arid
and semi-arid landscapes and occurs once deposition areas
become vegetated. The vegetation reduces flow velocity and
traps sediment so the deposition area acts as a bottleneck
restricting the amount of material transported downstream
and the distance sediment travels. Deposition areas also
have a greater infiltration capacity than other areas, being
relatively uncompacted, so less water flows downstream and
transport capacity is reduced. Buffering would not occur
without vegetation because deposition would eventually
increase local slopes to the extent that the higher sediment
transport capacity would prevent further deposition.
However, because areas of fill constantly trap soil, seeds,
nutrients and runoff water, so much vegetation grows that
they can no longer be removed. It is then common to find
deposits building up to the extent that flow is diverted
around them (Pickup, 1985).

B.   Regional Differences in STF Sequences

While the extent of erosion is a major variable
influencing the size and amplitude of STF sequences (see
next sub-section), other factors may play a part. If the
effect of these factors is randomly distributed across the
landscape, it can be treated as a noise term in the local
response model. If not, it may be necessary to develop a
trend model to describe any systematic regional variations
which exist. These variations arise for three main
reasons:

(1) the presence of different soil types;
(2) differences in the erosional process involved;
(3) regional variations in sediment availability.

No systematic survey has been carried out on how STF
sequences vary from one soil type to another. It does,
however, seem that coarse-grained material which has a high
infiltration capacity and a high critical velocity for
entrainment to occur, develops small, low amplitude
sequences compared with finer material such as clays. At
the same time, in arid landscapes, many differences in
surficial material are created by STF sequences themselves,
so there may be no need to differentiate between soil types.

Differences in the erosional process involved can
produce obvious contrasts in the type of STF sequence
present.   This is particularly true where a major floodplain
is bordered by terraces, large alluvial fans or slopes.   The
floodplain usually has very long sequences of high amplitude
whereas in the bordering areas, they are much shorter and
some of the deposition areas in the sequence may be missing.
    Changes in the pattern of such sequences arising
because of differences in soil type or process occupy
discrete regions in a landscape and can be handled by subdividing the system and applying different local response
models.   Differences in sediment availability, which arise
mainly due to grazing pressure, occur as gradual trends
and must be derived from models of how a landscape is used
by grazing animals (Pickup and Stafford Smith, 1987).
    In the arid zone, stock graze out from watering points
and their distribution is determined by a large number of
factors including distance from water, preferred vegetation
type for grazing, camping sites, wind direction, etc.
(Stafford Smith, in press).   In the case of cattle, the most
important of these are distance from water and vegetation
type (Hodder and Low, 1978).   The effect of grazing is to
reduce erosion resistance by removing protective vegetation
cover and to increase sediment detachment by trampling.
These activities are centred around watering points to which
the animals must return at regular intervals to drink.   In
a uniform landscape, this produces a radial pattern in which
the intensity of use decreases outwards from the water point
(Fig. 3).   When the landscape is not uniform, a more complex
pattern results (see, for example, Senft *et al*, 1983).   In
the case of cattle, where forage preference is a major factor
determining where the animals graze, unpalatable plant
communities close to water may be ignored for more palatable
ones further out.   Community preference can therefore
modify the radial structure by creating axes of concentration
or corridors as Figure 3 shows.
    The effects of grazing animals on a single STF sequence
is easy to predict.   Where heavy grazing or trampling occurs
over the whole of such a sequence, it will increase the
intensity of erosion in the scour zone, possibly convert
the transfer zone to an area of scour, and reduce the
propensity for deposition to occur.   It may even destroy
the sequence completely, replacing it with a larger one in
which deposition occurs at a greater distance from the
water point where the effects of grazing or trampling are
less extreme.   Alternatively, the STF sequence may be
preserved but with an increase in amplitude as more erosion

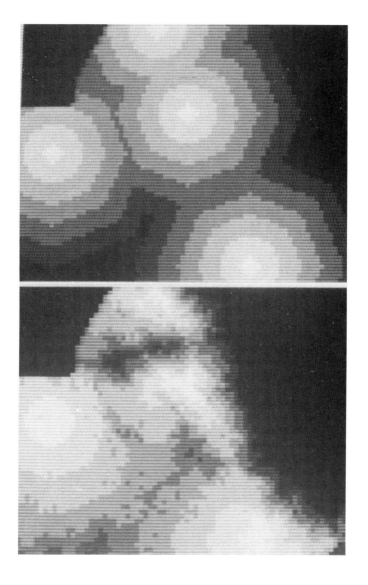

FIGURE 3.   *Idealised distribution of usage around four*
water points by grazing animals in a paddock with uniform
vegetation (upper image) and non-uniform vegetation (lower
image).   The level of usage decreases as the image tone
becomes darker.

produces more deposition in existing areas of fill in spite
of the fact that they are being grazed and trampled. This
response is particularly common in the larger sequences and
can often be recognised by population explosions of trees
and shrubs in areas of deposition in many parts of the arid
zone.

The effects of grazing on a complex mosaic of overlaid
STF sequences of different sizes is more difficult to predict.
Not all the sequences operate simultaneously or at the same
rates. Some may be active regularly, responding to minor
storms, whereas others only operate during very large, rare
events. Some may well disappear completely during extreme
events only to be recreated subsequently. The main effect
appears to be that large sequences are enhanced at the
expense of small ones as the next sub-section indicates.

C. *Regional Scale Landscape Change*

STF sequence behaviour and buffering provide the basis
for constructing models of soil erosion and deposition at
the regional scale. In a system with only limited erosion,
buffering restricts the length and amplitude of the STF
sequences. There is, therefore, a balance between the
forces which produce erosion and those which resist it. If
the intensity of erosion is increased or the erosion
resistance of the landscape is reduced, the smallest and
least persistent STF sequences are destroyed first and
become sediment source areas. A single large sediment
source area is likely to generate more runoff than a similar
area of many small ones so erosive forces increase. Sediment
can then only be trapped in major deposition areas because
the small ones have been destroyed by erosion and the system
is no longer buffered. A mosaic of small, low amplitude
STF sequences is therefore replaced with a few large ones,
in which large quantities of sediment move long distances
from sediment source areas to zones of fill. If a landscape
restabilises, the opposite occurs. As the forces of erosion
decrease or erosion resistance increases, sediment moves
a shorter distance from source areas and a new set of small
STF sequences develops. At first they may not persist but
eventually some survive, gradually increasing the level of
buffering in the system.

The changes described here are reflected by the
frequency distribution of sub-areas or cells within the
system in various states of erosion and deposition (Pickup,
in press). In a stable, buffered system, variance is small
and most points cluster round a central stable condition.

In an eroded unstable system, variance is higher with more points at the extremes of severe erosion and extensive deposition. There are also differences in the spatial auto-correlation function which decreases more rapidly with increasing spatial lag in stable buffered systems than in unstable ones. This reflects the change in the size and amplitude of the STF sequences present as the system changes.

Regional scale landscape change is an episodic rather than a gradual phenomenon and can occur quickly once initiated. The conditions for rapid change seem to be heavy grazing and trampling over large areas followed by one or more large rainfall events. The grazing removes vegetation, reducing the level of buffering, and the associated trampling detaches soil particles, making them available for entrainment by surface runoff. Whether that soil actually moves depends on the magnitude of the rainfall events which follow. If they are large enough to allow regrowth of vegetation and reattachment of particles to the main body of soil but not so large as to cause major runoff, the landscape rebuffers and no major changes occur. If a large rainfall event occurs, a substantial amount of sediment moves and the smallest STF sequences are replaced by larger ones. The extent to which this process continues depends on whether the initial heavy grazing continues or whether the landscape is allowed to rebuffer during sub-sequent smaller rainfall events. Rebuffering can only occur if seed banks survive, so the ability to rebuffer becomes progressively less as erosion increases. It is therefore likely that, as a major landscape change occurs, there is progressively less chance to reverse it.

D.   *Local Response Model - Implementation*

The local response model describes the interactions which occur between individual grid cells in the regional model and is a stochastic process capable of reproducing the combination of changes likely to occur in a STF sequence. These changes are shown in Figure 4 and can be individually stated as:

(1) a shift in the average level of scour and fill;
(2) an increase or decrease in the amplitude of scour or fill but no change in position;
(3) the spread or concentration of scour or fill up-stream or downstream.

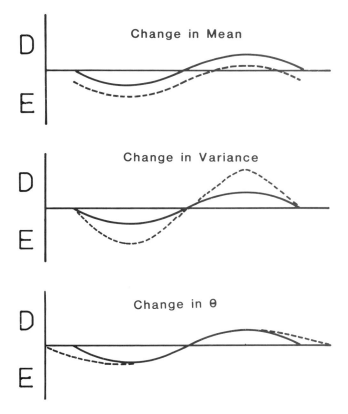

FIGURE 4. *Schematic diagram showing some of the changes in a STF sequence which may occur in one dimension. The upper graph denotes a change in the average amount of scour and fill. The middle graph indicates a change in the amplitude of the sequence producing a change in variance. The lower graph illustrates spread of the sequence and is reflected in a change in the θ values of the local response model used to forecast erosion (see below). On the vertical scale E denotes erosion and D indicates deposition.*

The changes are shown in one-dimensional form for simplicity but are, in fact, two dimensional and can be non-isotropic. In a real landscape, the problem is more complex because there is a mosaic of overlaid, interlocking STF sequences which interact with each other simultaneously. This situation is difficult to handle using conventional erosion models based on the continuity equation or the convection-diffusion equation. It can, however, be represented by stochastic process models with a variety of

forms. In this case, a simultaneous autoregressive random
field (SAR) model is used with the form:

$$Y(s) = \sum_{r \varepsilon N} \theta_r \, Y(s+r) + \sqrt{\rho w(s)} \qquad (1)$$

where Y is Pickup and Nelson's (1984) soil stability index
value at location (s), s refers to grid coordinates with
$(s = (i,j) \ \varepsilon \ \Omega)$, $\Omega = (i,j)$, $1 \leq i,j \leq M$ for a square data
matrix, r indicates the coordinates of the neighbourhood
set, N, around location s with $(r = (k,l) \ \varepsilon \ N)$; θ_r are the
weighting parameters; ρ is the variance of the noise series
and w(s) is a sequence of random variates with zero mean and
unit variance. Specifying location, s at coordinates 0,0
gives a first order neighbourhood set labelled as follows:

$$r = (0,1)$$

r = (-1,-1) x	x	x r = (1,-1)
r = (-1,0) x	.	x r = (1,0)
r = (-1,1) x	x	x r = (1,1)

$$r = (0,1)$$

SAR model parameters may be derived empirically from a given
data set using the fitting procedures developed by
Kashyap and Chellappa (1983). Their application to erosion
modelling is described by Pickup and Chewings (1986a,
(1986b), so only a brief account is presented here.
    The model expresses each value of the soil stability
index in an area as a weighted sum of the surrounding values
plus a noise component. The following description, while
not strictly mathematically correct, may help explain how
this structure can be used to generate the patterns of ·
erosion which occur when there is a regional scale landscape change. Firstly, consider a landscape without
significant erosion or deposition. STF sequences are only
present in incipient form and little structure is present.
This landscape may be represented by the noise series, w(s)
which consists of a set of soil stability index values
clustered about a mean representing a stable condition.
If erosion begins, there is more erosion and more deposition
so the spread of stability index values around the mean
increases. This spread is represented by the noise series

variance, ρ, which acts as scaling parameter. Erosion also results in the development of STF sequences as eroding areas expand and coalesce and sinks develop. The extent of this patterning is described by the θ parameters which increase as the STF sequences become larger.

There is, at present, no method for predicting how the θ and ρ parameters in the SAR model change with time as a system becomes more eroded or recovers. It is, however, possible to derive values for these parameters for areas of the same landscape which are in different states of erosion. This makes it possible to show how a particular area changes if erosion increases by using a similar but more eroded area as a prototype. The procedure is to derive estimates of the θ and ρ parameters for the prototype area and the area in which change is to be forecasted by fitting SAR models to each. The noise series values for the forecasting area are obtained by inverse filtering with SAR model parameters for that area. This noise series can then be used with the model parameters of the prototype area to generate a new set of Y(s) values. The result is that the statistical characteristics of the prototype area are transferred to produce a forecast but the spatial patterns in the forecasting area which are largely independent of erosion are maintained.

An example of the type of results which may be obtained using the forecasting procedure is presented in Figure 5. The forecast was made using a prototype from a more eroded area and shows how the eroded area expands and intensifies, thereby becoming more uniform. There is also an increase in the intensity of deposition although the area affected does not greatly increase. Both types of change accord with what might be expected in a real-world situation. More detail on model accuracy including a number of tests is presented by Pickup and Chewings (in press).

The erosion forecasting procedure relies on the assumption that latent or incipient STF sequences are present in the landscape and may be detected in Landsat data. This assumption is reasonable in many landscapes because of their geomorphic and climatic history. Virtually all arid zone landscapes contain features resulting from geomorphic processes which operated in the past but are less or no longer active today. Many of these features were associated with more active erosion and deposition although the reason for that activity is still a matter of speculation. When grazing animals produce erosion, they tend to reactivate the old patterns of activity which intensify and expand to occupy a larger proportion of the landscape. This makes it possible to use the latent pattern of erosion to forecast

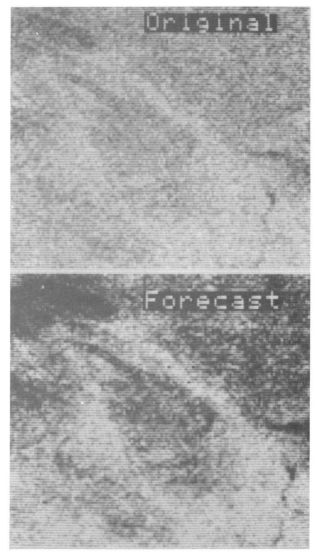

FIGURE 5.   Result of an erosion forecasting exercise.
The upper grey scale image represents the current state of
a 128x128 Landsat pixel area (approx. 74 km². The lower
grey scale image is its forecasted state using a prototype
from a more eroded area.   The light tones indicate erosion,
the dark tones indicate deposition while the intermediate
tones show areas of relative stability.

the new one. The method breaks down when a completely new
pattern develops which is unrelated to past conditions. This
sometimes occurs in the vicinity of waterpoints and makes
it necessary to add a trend model to the forecasting
procedure as described in the next section.

E. *Trend Model*

Attempts to predict observed change in landscapes
suggest that, in some cases, an adequate result can be
obtained from the local response model alone. In other cases
systematic regional errors occur. These errors can sometimes be reduced by dividing the landscape into segments
and using different local response models. If multiple
local response models do not work, a trend model should be
used. This introduces systematic regional changes in mean,
variance or both. It can also be used to insert new
erosion patterns in the landscape.

The trend model for cattle grazing effects is still
under development. Initial tests suggest that the
distribution of animals when grazing can be represented by
the convection-diffusion equation which may be written as:

$$\frac{1}{2} \sigma^2 \left(\frac{\partial^2 p}{\partial x^2}\right) - v \frac{\partial p}{\partial x} = \frac{\partial p}{\partial t} \tag{2}$$

where p is the density of animals at a distance, x from the
waterpoint, t is time, σ^2 is the variance parameter and v is
the drift. The solution of this equation for an input at
x = 0, t = 0 can be expressed as an inverse Gaussian
density function:

$$f(t,x,v,\sigma) = \frac{x}{\sigma(2\pi t^3)^{\frac{1}{2}}} \exp\left[-(x-vt)/2\sigma^2 t\right] \tag{3}$$

in which the parameters t, v and σ all vary with
vegetation type and reflect grazing preference.

The pattern of usage produced by this function for a
particular vegetation type is shown in Figure 6. Under
conditions of abundant forage, in stage 1 the number of

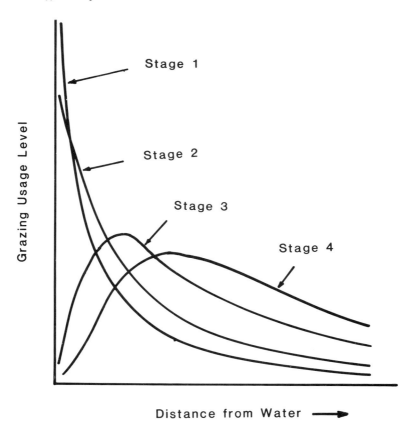

FIGURE 6. *Modelled intensity of grazing by cattle with increasing distance from water as the area is progressively eaten out.*

cattle grazing in a particular area declines exponentially with distance from water. Over time, as the area is grazed, the animals move progressively further out and the distribution becomes flatter as in stage 2. Eventually, the area closest to the water point becomes eaten out and is no longer used for grazing so the whole pattern is progressively displaced outwards as stages 3 and 4 indicate.

The inverse Gaussian density function only describes the location of grazing activities. Sediment availability is also related to the pattern of trampling which results from animals returning to the watering point to drink each day. The current procedure for obtaining the trampling pattern is to determine the shortest path between each grazing location and the water point and to sum the number of animals moving along those paths.

The derivation of trend models for landscapes with a number of plant communities present currently requires simulation. In this operation the location of grazing animals is calculated for each grid cell from equation (2) and the trampling pattern obtained in a separate operation by routing them to water step by step along the shortest path. A better approach might be to use the results of a simulation to obtain much simpler generalised relationships which can be fitted to observed trends. These trends may then be removed and replaced with others typical of greater or reduced stocking densities. Future work on this approach is planned.

V.   MODELLING AND RISK ASSESSMENT

The principal weakness of the regional models described above is that they can only describe the changes which occur when a landscape shifts from one state to another. They do not predict when that shift is likely to occur or the rate of change through time in response to external factors such as grazing intensity or rainfall. Rates of change are a difficult problem because the landscape does not respond uniformly and some parts of the landscape are affected by what is going on in neighbouring areas. A simple answer to these difficulties is therefore not expected.

These problems do not prevent the model from being used operationally for erosion risk assessment as long as enough prototypes are available for the local response model. These prototypes can easily be derived from archived LANDSAT data by a skilled interpreter. Thus data availability is not a problem. A risk assessment can then be carried out by forecasting landscape changes in an area for a variety of end conditions. Such forecasts make it possible to identify areas of continuing erosion, deposition and potential recovery. Once developed, the addition of trend models will improve the forecasts and will make it possible to identify the effects of changes in waterpoint location, fence lines, etc.

REFERENCES

Blong, R.J. (1966) Discontinuous gullies on the volcanic

plateau. *J. Hydrol. (N.Z.)*, 5, 87-99.

Borah, D.K., Alonso, C.V. and Prasad, S.N. (1982)   Routing graded sediments in streams: applications. *J. Hydraul. Div. Procs. Am. Soc. Civ. Engrs.*, 108, 1504-1517.

Box, G.E. and Jenkins, G.M. (1976)   *Time Series Analysis: Forecasting and Control*, Holden-Day, San Francisco.

Church, M. and Jones, D. (1982)   Channel bars in gravel-bed rivers.   In R.D. Hey, J.C. Bathurst and C.R. Thorne (eds.)   *Gravel-bed Rivers*, Wiley, Chichester, 291-338.

CSIRO and MPA (Australia) (1986)   *Micro-BRIAN: Software for Image Processing - User Manual.*   Microprocessor Applications Pty. Ltd., Melbourne.

Edwards, K. and Charman, P.E.V. (1980)   The future of soil loss prediction in Australia. *J. Soil Cons. NSW*, 36, 211-218.

Eyles, R.J. (1977)   Changes in drainage networks since 1820; Southern Tablelands, N.S.W. *Aust. Geogr.*, 13, 377-383.

Foster, G.R. (1982)   Modelling the erosion process.   In C.T. Haan (ed.)   Hydrologic Modelling of Small Watersheds. *Am. Soc. Agric. Engrs. Monograph*, 5, 297-379.

Higgins, R.J. (1976)   *Calibration and Testing of a Sediment Transport Model for the Kawerong-Jaba River.*   Report ED76/HY08 Bougainville Copper Ltd.

Hodder, R.M. and Low, W.A. (1978)   Cattle distribution of free ranging cattle at three sites in the Alice Springs District, Central Australia. *Australian Rangelands J.*, 1, 95-102.

Kashyap, R.L. and Chellappa, R. (1983)   Estimation and choice of neighbours in spatial interaction models of images. *IEEE Trans. on Information Theory*, 29, 60-72.

Leopold, L.B., Wolman, M.G. and Miller, J.P. (1964)   *Fluvial Processes in Geomorphology*, Freeman, San Francisco.

Meade, R.H. (1985)   Wavelike movement of bedload sediment, East Fork River, Wyoming. *Environ. Geol. Water Sci.*, 7, 215-225.

Pickup, G. (1974)   *Channel Adjustment to Changed Hydrologic Regime in the Cumberland Basin, New South Wales*, unpublished PhD thesis, Univ. Sydney.

Pickup, G. (1985)   The erosion cell - a geomorphic approach to landscape classification in range assessment. *Australian Rangelands J.*, 7, 114-121.

Pickup, G. (in press)   Hydrology and sediment models.   In M.G. Anderson (ed.), *Modelling Geomorphic Systems*, Wiley, Chichester.

Pickup, G. and Chewings, V.H. (1986a) Mapping and forecasting soil erosion patterns from Landsat on a microcomputer-based image processing facility. *Australian Rangelands J.*, 8, 57-62.

Pickup, G. and Chewings, V.H. (1986b) Random field modelling of spatial variations in erosion and deposition in flat alluvial landscapes in arid central Australia. *Ecol. Modelling*, 33, 269-296.

Pickup, G. and Chewings, V.H. (in press) Forecasting patterns of erosion in arid lands from Landsat MSS data. *Int. J. Remote Sensing*.

Pickup, G. and Nelson, D.J. (1984) Use of Landsat radiance parameters to distinguish soil erosion, stability and deposition in arid central Australia. *Remote Sensing Env.*, 16, 195-209.

Pickup, G. and Stafford Smith, M. (1987) Integrating models of soil dynamics, animal behaviour and vegetation response for the management of arid lands. *Aust. Geogr.*, 18, 19-22.

Rose, C.W., Williams, J.R., Sander, G.C. and Barry, D.A. (1983a) A mathematical model of soil erosion and deposition processes: I. Theory for a plane land element. *Soil Sci. Soc. Am. J.*, 47, 991-995.

Rose, C.W., Williams, J.R., Sander, G.C. and Barry, D.A. (1983b) A mathematical model of soil erosion and deposition processes: II. Application to data from an arid zone catchment. *Soil Sci. Soc. Am. J.*, 47, 996-1000.

Rieger, W.A., Olive, L.J. and Burgess, J.S. (1982) The behaviour of sediment concentrations and solute concentrations in small forested catchments. *Inst. Engrs. Aust. National Conference Publication*, 82/6, 79-83.

Schumm, S.A. and Hadley, R.F. (1957) Arroyos and the semiarid cycle of erosion. *Am. J. Sci.*, 255, 161-174.

Senft, R.L., Rittenhouse, L.R. and Woodmansee, R.G. (1983) The use of regression models to predict spatial patterns of cattle behaviour. *J. Range Management*, 36, 553-557.

Stafford Smith, M. (in press) Modelling: three approaches to predicting how herbivore impact is distributed in rangelands. *US Dept. Agric. Tech. Memo*.

Wischmeier, W.M. and Smith, D.D. (1978) Predicting rainfall erosion losses - a guide to conservation planning. *U.S.D.A. Agric. Handbook*, 537.

Woods, L.E. (1983) *Land Degradation in Australia*. Australian Govt. Printing Service, Canberra.

7
Stratigraphic Complexities and the Interpretation of Alluvial Sequences in Western Australia

Karl-Heinz Wyrwoll

Department of Geography
University of Western Australia
Nedlands, WA

Und sehe, dass wir nicht wissen können!
Das will mir schier das Herz verbrennen.
GOETHE, Faust.1

I. INTRODUCTION

Quaternary alluvial deposits are strong elements in the geomorphology of many areas of Western Australia, and have proven to be of considerable palaeoenvironmental and archaeological importance (Merrilees, 1968; Wyrwoll, 1977; 1984; 1988; Wyrwoll and Dortch, 1978; Bordes *et al*, 1980; Pearce and Barbetti, 1981). An inevitable requirement which has to be met in their study is the development of a secure stratigraphic framework - with reliable absolute age control. While adequate stratigraphic control is desirable for any sediment sequence, this task is especially vexing for Quaternary deposits because of the high degree of time resolution required and the stratigraphic complexities that alluvial deposits exhibit.

The need for high resolution stratigraphy is vital when alluvial deposits are used in palaeoenvironmental/palaeoclimatic reconstruction, where the identification of relatively short-lived events needs to be achieved. While the use of Quaternary alluvial deposits in palaeoenviron-

Copyright © 1988 by Academic Press Australia.
All rights of reproduction in any form reserved.

mental reconstruction has often been viewed critically (eg
Bowler *et al,* 1976), and despite the problems encountered
in their use, alluvial sequences do provide a great deal
of useful palaeoenvironmental information. They are
especially important for the geomorphologist as they
register periods/events of geomorphological activity,
whereas other sediment and/or biological sequences are
frequently more accurate recorders of 'mean' climatic/
environmental states. But the need for a secure stratigraphy
is no less important for geoarchaeological work, where it
may be imperative to have sound stratigraphic control to
evaluate the significance of sites of archaeological
associations.

   In the case of many alluvial sequences, exposure is
restricted to a limited number of sections, few of which
can be dated because of prohibitive costs or the absence
of suitable material. Nevertheless,such limited data are
often used to make stratigraphic generalizations. Often
the stratigraphic complexity of alluvial deposits only
becomes apparent when alluvial sequences are 'walked out'
- which is seldom possible. Without this, considerable
care has to be taken that the sections available for study
adequately capture the variability of the alluvial
succession. Only then, and this is left to the subjective
judgement of the individual concerned, can the data obtained
be considered to be representative of drainage basin
behaviour, rather than being indicative of more local
controls. This is a fundamental consideration when a study
aims to identify hydrological/sediment yield changes at a
regional scale. For despite the difficulties and
ambiguities of this type of work, it remains a primary
purpose of many studies of alluvial deposits, and is
facilitated by a large body of work (Schumm, 1977; Hickin,
1983; Gregory, 1983; Bowler, 1986; Pickup, 1986).

   This essay draws attention to the difficulties that have
been encountered in working on the stratigraphies of three
Quaternary alluvial sequences in western Australia. The
problems are presented at three spatial scales: (1) in a
channel reach, (2) in a wider transitional geomorphological
setting, where a river leaves a gorge section and enters
a coastal plain and (3) at the overall scale of a drainage
basin. In part an attempt is made to outline the
theoretical context, almost in the form of a hypothesis,
in which the field evidence can be placed. Although the
link between the theoretical basis and the field evidence
is tentative, it is hoped that combined they reinforce the
need to consider the possible stratigraphic complexities

of Quaternary alluvial deposits before drawing regional
stratigraphic/hydrological/sediment yield inferences.
Locations referred to in the text are shown in Figure 1.

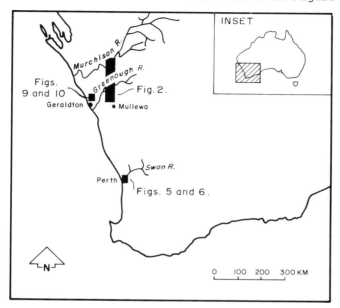

*FIGURE 1. Locations referred to in the text and on
subsequent figures.*

II.   CHANNEL REACH AGGRADATION AND RESULTANT STRATIGRAPHY

Changes in channel depth and/or slope along a reach may
occur purely as the result of local factors. The
aggradational response that such hydraulic changes force
on the stream channel are accommodated in a wide range of
facies of which stream bars are important. These provide
the major mechanisms for channel adjustment to changes in
the sediment transport capacity of a channel reach. So
clearly, local hydraulic changes can invoke direct
sedimentological responses. In these bars, as channel
macroforms which have a high preservation potential, they have
a good chance of being registered in the stratigraphic
record. When such changes are integrated in time within
an alluvial fill, they can give rise to a very complex and
diachronous stratigraphy.

From general geomorphological considerations it would

appear that localised stream aggradation events should be
important along streams which transport limited amounts of
bedload, possess a complex planform with bedrock-alluvial
reaches, and where they are set in large drainage basins
with limited relief differences, in which sediment is contributed from areas throughout the basin and where the basin
has a low sediment delivery ratio. These attributes
describe the larger streams of the Murchison District of
the Yilgarn Block of Western Australia. In this region
alluvial deposits, although seldom more than 4 m thick, are
prominent along the main channels of the basins. These
deposits are well exposed and it is possible to walk out
large stretches of sections, making it feasible to work out
the details of the alluvial stratigraphy.

The "Murchison Cement" is an informal stratigraphic
name first proposed by Merrilees (1968) for the alluvial
sequences of this area. The deposits provided some of the
first evidence in Australia of the association of an extinct
diprotodontid *(Zygomaturus trilobus)* with stone artifacts.
This and other evidence allowed Merrilees to develop his
discussion of the role that man may have played in causing
the extinction of the Australia megafauna - a theme which
is still receiving much attention (articles in: Martin and
Klein, 1985). During the course of initial work (Wyrwoll
and Dortch, 1978), the Murchison Cement was studied along
the Greenough River north of Mullewa (Fig. 2). A working
stratigraphy was set up which divided the alluvial series
into an 'Older Fill' and a 'Younger Fill'; the two being
separated by a strong pedocalcic horizon.

The 'Older Fill' is a moderately indurated, in part well
cemented, alluvial complex which attains a thickness of up
to 4 m. The sediments are variable in texture but with a
predominance of material in the medium to coarse sand size
fraction. Primary sedimentary structures are generally
poorly preserved. Lower flow regime bedform structures are
evident in some of the coarser sand horizons. The
sediments are predominantly channel deposits. Carbonate
segregations, calcrete, partly silicified calcrete and
associated nodules and bands of opaline silica occur
extensively throughout this unit. So far it has not been
possible to date alluvial units of the 'Older Fill'.

The grain size characteristics of the overlying 'Younger
Fill' complex are relatively uniform, being largely medium-
fine sand and silt, and were deposited largely as overbank
sediments. The sediments are slightly compacted and
earthy carbonate segregations occur throughout; this contrasts sharply with the hard, silicified calcrete

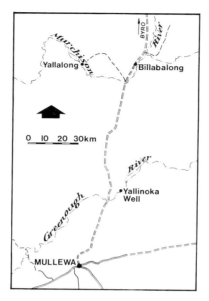

FIGURE 2. *Location of the sections shown in Figures 3 and 4. The Greenough River sections extend from Yallinoka Well to approximately 5 km west of the road. Murchison River section G.39, is located at Yallalong, the other four sections extend from approximately 1 km west of the road to about 3 km north-east of Billabalong.*

horizons present in the 'Older Fill'.

Later, the details of the stratigraphy of the deposits along some 15 km of the middle section of the Greenough River were established (Fig. 3). The stratigraphic marker unit was the strong pedocalcic horizon which was originally used to define the disconformity between the 'Older' and the 'Younger' fills. In the older, three alluvial units predate the development of this horizon (Fig. 3). The oldest of these (only shown in section G.22) is an extensively weathered ferruginized fill, with weak ferricrete development. Fill F is a distinctive alluvial complex which has been extensively silicified and silica cemented (Fig. 3). Its sediment characteristics are variable, with the primary depositional structures largely destroyed. It has a characteristic green-red colour with the sediments having a clayey matrix. Overlying F in some sections, is a less altered alluvial fill, in which the pedocalcic horizon (G.23, G.26, etc.) has its strongest expression. The correlations which have been attempted between sections are

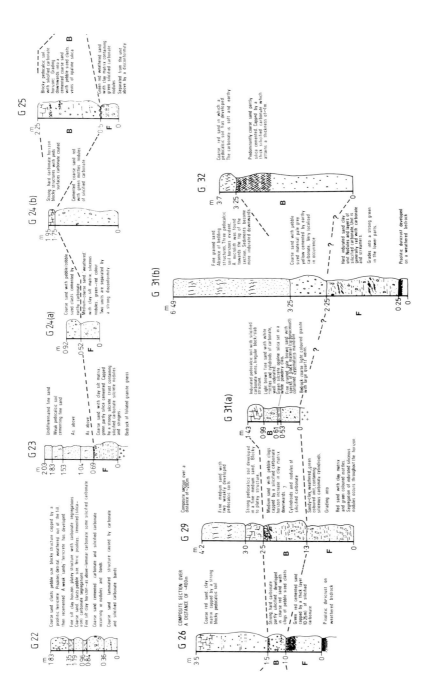

FIGURE 3. Details of the alluvial stratigraphy in the middle reaches of the Greenough River
— see Figure 2 for location.

clearly tentative, and based on lithostratigraphy. There is no reason to assume that a time-based stratigraphy would conform to this tripartite division. At best the stratigraphy of the Greenough can be seen as one which gives some recognition to the spatial and temporal variability which is an inherent feature of these deposits. This conclusion is just as applicable to the channel sequences of the alluvial succession of the middle Murchison River - the type area of the Murchison Cement (Fig. 4).

These field sections show that the channel sequences of the Murchison and Greenough alluvial deposits, when traced even over limited distances, consist of a number of alluvial units, some of which are likely to be specific to one section. Correlation of any chronological framework can only be resolved by a large number of absolute dates.

For the purpose of discussion, local controls of channel aggradation have been emphasized but clearly other factors need to be considered. For instance, the effects of channel avulsion and migration on alluvial architecture must be considered. The problem of stratigraphic variability is further shown in the 'Younger Fill', and because pedogenesis and weathering are more subdued in these units, it is easier to identify individual aggradation events (Figs. 3 and 4). It is evident that the deposition of a particular unit of the 'Younger Fill' was again quite often a localized event; this is an expected feature of overbank deposition.

From these findings and the complexity demonstrated, it is clear that the statement of Vita-Finzi's (1973, xi) that: "...as the work proceeded, I became convinced that dating, rather than serve as an adjunct to stratigraphy, deserved to replace it as the basis of chronology", is of great relevance to this study.

However, this may not be possible because of the absence of dateable material, errors inherent in the various dating techniques, the nature of the material used for dating (possible contamination by younger/older material), the impracticality of a large number of dates and the problem of having enough sections to capture the variability of an alluvial fill complex.

III. WIDER GEOMORPHOLOGICAL CONTROLS ON ALLUVIAL
    STRATIGRAPHY

The Swan River alluvial deposits provide the example of the next scale - that of a transitional geomorphological

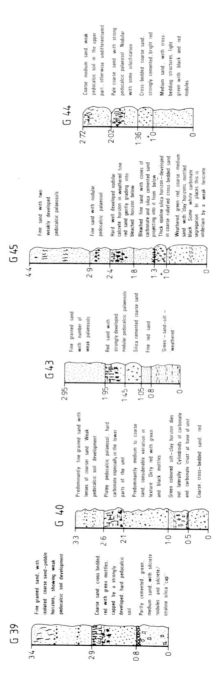

FIGURE 4. Alluvial stratigraphy of the middle reaches of the Murchison River - see Figure 2 for location.

setting. The Swan is the major river in southwestern Australia, and in recent years has attained some prominence in Australian Quaternary studies, because aboriginal stone artifacts, dated at around 39K years BP, have been found in its alluvial sequences (Pearce and Barbetti, 1981). These are among the oldest assemblages so far found in Australia.

The river leaves the Darling Scarp some 25 km NNE of Perth. There it has developed a broad alluvial plain with well developed terraces (Aurousseau and Budge, 1921) (Fig. 5). This area is clearly dominated by a 20 m terrace whose deposits yielded the artifacts. More recently other stone artifacts have been found in this terrace nearer to the scarp at the 'Terrace Site' (Fig. 5), by the author and M.L. Schwede. This prompted a wider study of the alluvial stratigraphy of the area.

A complex alluvial stratigraphy is expected in this kind of geomorphological setting. Once the stream leaves the gorge confines of the Darling Scarp aggradation, planform adjustment and stream avulsion would have been likely. This kind of response is well illustrated by Fahnestock and Bradley (1974) and Hong and Davies (1979). With stream avulsion accompanied by mobile-belt migration, and with aggradation punctuated by erosional events, a complex alluvial stratigraphy must inevitably result downstream from the scarp. This can be illustrated by a simple random walk model, which is taken as being analogous to stream avulsion.

Figure 6 shows a series of random walks generated on a grid by selecting random numbers to determine the direction of movement from one grid square to the next. It was specified that the walk may move to adjacent grid squares but cannot move backwards. This simple scheme, developed by Leopold and Langbein (1962) is used only for illustration. A stream emerging from a restricted gorge section undergoes a change in the degrees of freedom it has in adjusting to its hydraulic constraints. In the diagram this is indicated by the core section emanating from the point of origin. The inclination of the sides, and hence the distance until a full channel belt is established, was arbitrarily chosen. In actual streams this distance should scale with stream power. The sides of the cone section and those of the mobile belt were used as barriers to channel migration. If a random number required that these barriers be crossed, this was ignored and another number generated.

From these results it is clear that a complex stratigraphy is to be expected in this area. Once the free channel belt is more established, the overall stratigraphy of the alluvial sequence becomes more simple.

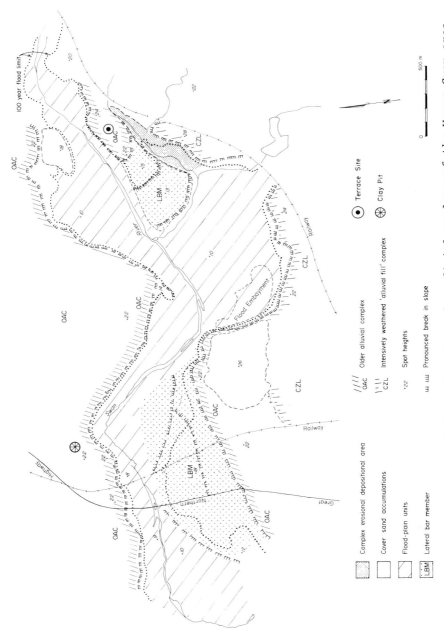

FIGURE 5. The major geomorphological units and surficial geology of the Upper Swan area.

GORGE SECTION

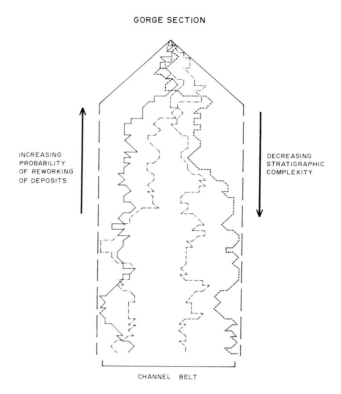

INCREASING
PROBABILITY
OF REWORKING
OF DEPOSITS

DECREASING
STRATIGRAPHIC
COMPLEXITY

CHANNEL BELT

FIGURE 6. A simple random walk model of channel migration downstream of a gorge section.

Exposures for study were limited to two major areas of sections, and these, in combination with overall surface outcrops, allowed the stratigraphy to be resolved at three levels of detail:

A. *The Initial Division of the Terrace into Two Alluvial Complexes*

From surface outcrop the terrace can be divided into two alluvial complexes: an 'Intensively Weathered Alluvial Fill Complex' and the 'Older Alluvial Complex' (Fig. 5). In order to distinguish these with confidence, a detailed examination of the pedological/weathering characteristics was undertaken.

The intensively weathered fill is indurated, weathered (as its name suggests) and leached giving rise to a 'kaolinized' appearance in outcrop. Exposure is poor, and because of this and the high degree of weathering, little can be said about the primary sedimentology of the unit. The sediments are variable in texture with both arenaceous and argillaceous members well represented. Some con- glomeratic lenses are also present. With the exception of the bedding structures preserved in these rudaceous beds, there has been a general loss of the primary structures. The coarse textured horizons indicate traction clogs and have suggestions of planar and low-angle planar crossbedding. The intensity of weathering and restricted outcrop makes it impossible to determine alluvial architecture.

The fill is capped by a ferricrete, and the nature of this is dependent on the texture of the host sediments. On clay rich units, strong pisolite development is evident; in more arenaceous units ferruginous 'crusts' are more likely. Ferricrete horizons also occur within the body of this alluvial complex. Ferrallitic mottling can be seen throughout, gley colourations are also present. Silicified - partly opaline - rhizoconcretionary structures are found in some parts. The sediments of this intensively weathered alluvial fill can be confidently distinguished from the younger complex by their distinctive micromorphological characteristics. In thin-sections, the skeletal framework is dominated by quartz, with an absence of other, more easily weathered silicates. The sediments show strong plasma development, with an almost isotic fabric (terminology after Brewer, 1976). 'Ghost' embedded cutan structures are evident, with indications of former mosepic plasmic fabric. In plane polarized light, channel develop- ment and plasma flow structures can be seen, with the plasma having a deep brown-ferric impregnation. Compound skew- plane ferri-argillans are present. Large, strongly separated nodules with undifferentiated fabric occur frequently. Also present are strongly adhesive, diffuse and irregular sesquioxide nodules with undifferentiated fabric.

The overall lateral extent of the intensively weathered alluvial complex is unclear, but sufficient indications are present to suggest that alluvial facies variations occur and that colluvial additions are likely. Topographically the weathered alluvial complex is continuous with the over- lying 'Older Alluvial Complex', and care has to be taken to separate the two complexes.

B. *Indications of Stratigraphic Complexity in the Older Alluvial Complex*

The geomorphological expression of the 'Older Alluvial Complex' is most evident in the north-east section of the area where a well defined terrace occurs, which has been dissected by a tributary on its eastern side, and eroded and partly covered by a younger set of deposits along its western extension (the 'Lateral Bar Member' shown in Fig. 5). The overall stratigraphy and sedimentology of this alluvial complex were examined in sections along the northern edge of the terrace at 'Terrace Site' (Fig. 5), where the artifact material was found.

Three major alluvial units could be separated in these sections:

(1) A basal unit in which the sediments are moderately indurated and extensively rubefied. However, both reducing and hydration colours are also present as mottles and stringers. In texture, the sediments vary from medium-fine sand to horizons with coarse sand containing clasts of pebble-sized material. No primary clay units are present. It is evident from thin-sections that the clay matrix, which occurs throughout the sediment pile, and which is the source of the red pigmentation of the unit, is due to post-depositional pedogenic-diagenetic alteration. The plasma is strongly developed with domains which show flecked orientation - sepic plasmic fabric. Embedded ferri-argillans are common. Vugh and channel argillans are also present. The argillans are strongly orientated and well separated. Orthoclase feldspars are strongly weathered, while the plagioclase feldspars are less weathered. Microcline grains are weakly weathered with pocked marks and weathering along crystallographic planes.

(2) The overlying unit is a yellow, fine-medium sand, which is moderately indurated in its upper part. Pedological differentiation in this unit is weak. In thin-section the sediments possess a strong skeletal grain framework - 55% angular grains, with a strong dominance of quartz. Minor feldspars: plagioclase, microcline, orthoclase and perthite are present, and these are only slightly weathered. Isolated lithoclasts can also be seen. Only weak-modern plasma segregations have developed, with some free-grained and embedded argillans, which show no orientation. There is a tendency towards the development of an argillic-sepic fabric.

Some textural contrast is present, with the unit grading downwards into a fine-medium more clay rich sand. Some

colour differentiation is evident, with the unit becoming
more red in colour down-profile, this is accompanied by an
increase in the occurrence of strong deep-red mottles. The
unit becomes loose and unconsolidated towards the base.
Soft to moderately hard carbonate nodules are present in
its lowest parts.

(3) The uppermost unit of this succession is a clay
rich red-yellow sand which reaches a thickness of 1.6 m.
A distinct textural contrast is evident in this unit, and
is the result of pedological differentiation. The lower
part of the unit is a fine-medium sand, with a sandy fabric.
It passes upwards into a coherent argillic horizon which
has a rough-ped to earthy fabric. The textural changes are
accompanied by changes in colour from a dark yellow-brown
to a red with yellow-grey mottling. Well formed deep red
sesquioxide pisolites (<10 mm diameter) are evident throughout much of the profile.

The upper unit has a significantly higher clay content
than the underlying unit, attaining maximum values of 42%
in the area of the argillic horizon. The textural
variability of the upper unit is both a primary sedimento-
logical feature (for instance, a 'clay-ball' was found in
the lower parts of the upper unit) and the result of
subsequent pedological modification. This conclusion is
apparent in thin-sections. In the argillic horizon strong
plasma development has occurred, with the plasma having an
insepic fabric grading to lattisepic fabric. Embedded
argillans are strongly developed in parts. Vugh/channel
argillans with strong orientation are present. A variety
of felspars are evident in thin-section. These are
generally slightly to moderately weathered. A number of
plagioclase felspar grains are strongly weathered.

The upper two units are essentially massive with a total
absence of bedding structures. From this, and the grain-
size characteristics of the sediments, it would seem likely
that these two units were deposited as overbank sediments.
This contrasts with the basal unit which is clearly a channel
facies. The disconformity separating the upper two units
is not marked by any evident pedological modification and
consequently is interpreted as marking a short period of
time. On the basis of the micromorphological work, which
has shown the development of strong pedological features
in the sediments of the basal channel unit, and the intensity
of felspar weathering, it is inferred that a strong disconformity exists between the channel unit and the overlying
sediments.

C. Alluvial Architecture of the Terrace Fill

The exposures so far considered have only given indications that the stratigraphy of the terrace fill may be quite complex. An indication of the degree of stratigraphic complexity that the terrace fill can attain in any one 'complete' section, is suggested in a 'Clay Pit' located further downstream (Fig. 5). In this section (Fig. 7), the various units of the terrace fill can be separated solely from their expression in the section, and there is no need to resort to micromorphological work.

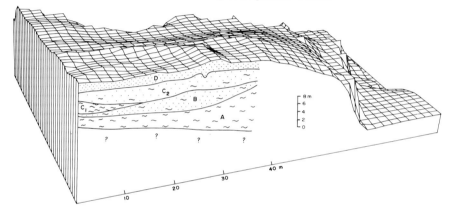

FIGURE 7. The stratigraphy of the terrace fill exposed in the 'Clay Pit' - see Figure 5 for location (approximate scale only).

The various units have undergone considerable pedological/weathering modifications which make it difficult to establish their primary sedimentology. The basal unit is a 'clay' (A) in which weathering and diagenetic modifications have been especially strong, and this makes it impossible to resolve the details of the stratigraphy. The unit has a general grey-green colour with strong sesquioxide mottling which grades into glaebule segregations; these are especially prominent in the upper part of the unit. Although the unit has a dominant clay texture, there are variations in the textural characteristics of the unit, with some parts having a significant coarse sand fraction. Accompanying the textural variations, are indications of lenticular and tabular beds. While these are difficult to identify, they suggest that the basal 'unit' should be further subdivided, but the sediments are so altered that this is not possible.

The basal 'clay' is overlain by a lenticular channel
fill (B), which is continuous over much of the section. The
unit is clay rich with coarse sand beds, containing pebble-
sized material in its upper parts. Some indications of
planar crossbedding occurs in the coarser horizons. Partly
silicified rhizoconcretionary forms and weak sesquioxide
glaebule development are evident in the upper part of the
unit.

The unit is in turn overlain by two channel fills. C_1 is
predominantly a clay-textured fill, while C_2 is more
variable in texture ranging from a coarse sand to sandy
clay, and in parts it is totally unconsolidated. In other
parts of the section it is a hard indurated mudstone. Low
angle planar bedding and planar cross bedding are apparent
in the coarser sand horizons. The upper part of the two
units are often indurated by platy carbonate which exploits
bedding surfaces. Rhizonconcretionary forms are also
present. Section units are separated by well marked dis-
conformities. This is not the case for the two fills (C_1 and
C_2).

The contact between the two C units and the uppermost
unit (D) is irregular and marked by small erosional channel
infills. D is a fine to medium pale yellow sand which has
a significant clay content. There are no apparent
sedimentary structures. Sesquioxide glaebules have developed
to variable degrees throughout the profile. These are
moderately indurated; more intensively indurated detrital
ferruginous pisolites are also present. In some areas of
the section there is a colour change down the profile with
a bleached horizon developing in the lower parts; this is
accompanied by the development of deep-red mottles. Bleaching
of the sediments is also evident around root channels. The
overall characteristics of this upper unit are consistent
with overbank origins. In more disturbed parts of the
section (not shown in Fig. 7), the uppermost unit pinches
out, and the limb of an underlying channel crops out on the
surface of the terrace.

With no dates, the only correlation between the 'Clay
Pit' and the 'Terrace Site' is between the uppermost units
at each site. However, this is based on general
lithostratigraphy, more specifically texture and inferred
mode of deposition. The pedological characteristics of the
two units are quite different; the argillic horizon present
at one site is absent at the other. This need not be of
any real chronological significance, and may just reflect
pedofacies variation. But from the available evidence it
is probably realistic to conclude that there is no real basis
for any temporal correlation.

Even in the short distance for which the alluvial stratigraphy has been considered, a complex picture of the alluvial architecture has become apparent. This supports the suggestions of the random walk model that alluvial aggradation was complex. Stream aggradation was clearly a protracted and discontinuous process, with respective fills/units not necessarily represented in each section. The complexities of the alluvial stratigraphy preclude any direct time correlation between sections, unless controlled by radiometric dates. The controls on local-scale channel aggradation events, should also not be forgotten. It is also evident that attempts to appeal to generalized external, climatic-base level controls as guidelines in chronostratigraphic correlation or stratigraphic inter-pretation are ill-advised.

IV. DIFFUSION OF AN ALLUVIAL FILL THROUGH A BASIN

The relationship between an alluvial fill and its sediment source areas is often obscure. This is especially the case in large basins, but in 'controlled' smaller basins, it is sometimes possible to identify sediment source areas. In these areas it is possible to evaluate 'threshold controls' on sediment supply, and use the results as a further step towards determining the wider environmental controls on sediment supply. However in those instances, where there is a direct association between a source area and an alluvial fill, stratigraphic implications also need to be considered. In a number of small drainage basins in the Moresby Ranges of the Geraldton area, it is possible to establish the Late Pleistocene alluvial fill source areas. The details of the stratigraphy and sedimentology of the fill, and the causes of stream aggradation are discussed elsewhere (Wyrwoll, 1984; 1988). Here, the role of alluvial fill is only considered in providing the setting for further stratigraphic considerations.

The drainage basins of the western side of the Moresby Range are small, with areas of 60-80 km^2. They have restricted long axes, have relatively large relief (about 200 m) and have simple drainage networks. The steep hill-sides of the escarpments bounding the basins provide sensitive sediment source areas (Fig. 8). These are the necessary attributes for a basin to be responsive in its sediment yield characteristics to changing environmental conditions. They also allow the sediment yield responses to be registered in the alluvial record.

BULLER BASIN

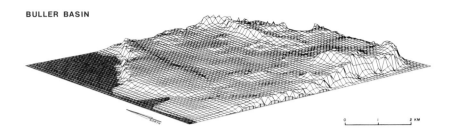

FIGURE 8.   Relief characteristics of the Buller basins in the Moresby Range (vertical exaggeration x 4).

It is clear from the location and geometry of the Late Pleistocene fill in the Buller Basin (Fig. 9) that, in the upper parts of the basin, stream aggradation was linked to high sediment supply originating from the surrounding steep hillslopes.   In such a location, where a stream builds up an alluvial fill downstream from the sediment source area, an aggradation front will move down the channel. It can be shown (Soni *et al,* 1980; Gill, 1983, Torres and Jain, 1984; Ribberrink and van der Sande, 1985) that the sediment continuity equation (1) can be used to develop a model which described this process.   The continuity equation takes the form:

$$\frac{\partial z}{\partial t} + \frac{\partial q_b}{\partial x} = 0 \qquad\qquad (1)$$

where z is bed elevation, t is the time, $q_{b'}$ is the volume per unit width per unit time of bedload discharge and x is the distance downstream.   This is then used to derive the parabolic partial differential equation:

$$\frac{\partial z}{\partial t} = K_O \frac{\partial^2 z}{\partial x^2} \qquad\qquad (2)$$

in which K_O is a diffusion coefficient.   With given boundary conditions equation 2 can be solved to describe the build-up geometry of an aggrading front.   Flume studies and

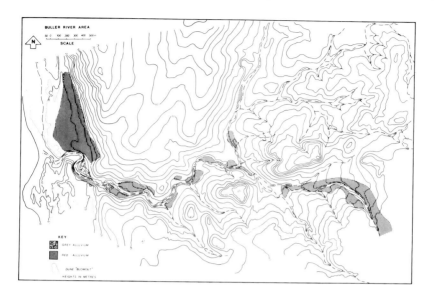

FIGURE 9. Quaternary alluvial fills in the Buller Basin. The 'Red Alluvium' is the Late Pleistocene fill discussed in the text.

theoretical work have shown that there is some value in seeing the aggradation process in this way. Although these studies have been restricted to laboratory and engineering time scales (but see Culling, 1960), this work has important implications for studies of stream aggradation over geological time by focusing on the use of the diffusion coefficient K_O.

In the Moresby Range alluvial sequences, the aggradation front has, in some basins, moved over distances of 4-5 km. The time taken to cover this distance is determined by the magnitude of K_O. It can be shown that with a constant sediment supply:

$$K_O \sim \frac{G_O}{S_O} \qquad (3)$$

where S_O is the initial channel slope and G_O is the transport capacity of the flow (de Vries, 1973; Soni *et al*, 1980). This equation gives some indication of the parameters which can determine the value of K_O. Clearly, only crude indications of these parameters could ever be obtained for stratigraphic sequences. But if the necessary resources were available it would be possible to obtain indications of the age variations along the aggradation wedge.

This model of stream aggradation cannot be placed in any real sense within a geomorphological - stratigraphic context. The aim was to show that theory could be used as a conceptual device, rather than a rigorous deductive guide. This model of stream aggradation has raised the question that the alluvial fill is time variant. It may not be possible to solve this but the point that is being made is that, if a high degree of time resolution is required, then such considerations need to be borne in mind.

V.   CONCLUSIONS

In emphasizing the problems of alluvial stratigraphies, this essay has taken a rather negative view of the complexities involved. These problems have been persistent in the study of Late Cenozoic alluvial deposits in Western Australia. These sequences are far removed from the ideal models of alluvial architecture - models which are inevitably set in subsiding basins with thick sediment sequences. In Western Australia, alluvial sequences are often poorly exposed, are found in areas which have seen no significant tectonic activity since at least the Proterozoic, where 'end-plains' of erosion dominate the regional geomorphology, and where it is possible to cross from the Holocene to the Late Tertiary over a lateral distance of 20 m in an alluvial deposit which is 4 m thick. Perhaps a geomorphologist could therefore be excused for developing an obsession with the problems of stratigraphy.

The recent work of Young *et al* (1986) and Waters (1985; 1986) has again shown the problems that are encountered in identifying and evaluating stream aggradational events at a regional scale. This essay has tried to highlight the stratigraphic problems associated with alluvial sequences when these are considered in channel reaches, in specific localities and small basins. The problems that are found at these three scales cannot simply be dismissed in an arm-waving appeal to "lumpers and splitters". The difficulties are inherent in the nature of alluvial deposits when viewed with the time resolution required of them in Quaternary work, and that for many sequences this may be an intractable problem.

ACKNOWLEDGEMENTS

I thank David Scott, Karen Wyrwoll, Bob Loughran and Rob Warner for their help. I am grateful to Buddy Wheaton for help with fieldwork.

REFERENCES

Aurousseau, M. and Budge, E.A. (1921) The terraces of the Swan and Helena Rivers and their bearing on recent displacements of the strand line. *J. Roy. Soc. WA.*, 7, 24-43.

Bordes, F., Dortch, C., Raynal, J-P. and Thibault, C. (1980) Quaternaire et Prehistoire dans le bassins de la Murchison (Australie occidentale). *C.R. Acadm. Sc. Paris*, t. 291, Serie D, 39-42.

Bowler, J.M. (1986) Quaternary landform evolution. In D.N. Jeans (ed.), *Australia: a Geography, Volume 1, The Natural Environment*, Sydney University Press, 117-147.

Bowler, J.M., Hope, G.S., Jennings, J.N., Singh, G. and Walker, D. (1976) Late Quaternary climates of Australia and New Guinea. *Quat. Res.*, 6, 359-394.

Brewer, R. (1976) *Fabric and Mineral Analysis of Soils*, Kreiger, New York.

Culling, W.E.H. (1960) Analytical theory of erosion. *J. Geol.*, 68, 336-344.

Fahnestock, R.K. and Bradley, W.C. (1974) Knik and Matunuska Rivers, Alaska: a contrast in braiding. In M. Morisawa (ed.) *Fluvial Geomorphology*, Binghampton State Univ., 221-250.

Gregory, K.J. (ed.) (1983) *Background to Palaeohydrology*, Wiley, London.

Gill, M.A. (1983) Diffusion model for aggrading channels. *J. Hydraul. Res.*, 21, 357-367.

Hong, Le Ba and Davies, R.R.H. (1979) A study of stream braiding. *Geol. Soc. Amer. Bull.*, 90, 1839-1859.

Hickin, E.J. (1983) River channels changes: retrospect and prospect. *Sedimentol. Spec. Publ.*, 6, 61-83.

Leopold, L.B. and Langbein, W.B. (1962) The concept of entropy in landscape evolution. *U.S.G.S. Prof. Paper 500-A*.

Martin, P.S. and Klein, R.G. (eds.) (1985) *Quaternary Extinctions: a prehistorical revolution.* Univ. of Arizona Press, Tuscon.

Merrilees, D. (1968) Man the destroyer: late Quaternary
    changes in the Australian marsupial fauna, *J. Roy.
    Soc. WA,* 51, 1-24.
Pearce, R.H. and Barbetti, M. (1981) A 38,000 year old
    archaeological site at Upper Swan, Western Australia.
    Archaeol. in Oceania, 16, 173-178.
Pickup, G. (1986) Fluvial landforms. In D.N. Jeans (ed.)
    *Australia: a Geography, Volume 1: The Natural
    Environment,* Sydney Univ. Press, 148-179.
Ribberrink, J.S. and van der Sande, J.T.M. (1985)
    Aggradation in rivers due to overloading - analytical
    approaches. *J. Hydraul. Res.,* 23, 273-283.
Schumm, S.A. (1977) *The Fluvial System,* Wiley, New York.
Soni, J.P., Garde, R.J. and Range Raju, K.G. (1980)
    Aggradation in streams due to overloading. *J. Hydraul.
    Div. ASCE,* 106, No. HY1, 117-132.
Torres, W.F. and Jain, S.C. (1984) Aggradation and
    degradation of alluvial-channel beds. *Iowa Inst. of
    Hydraul. Res.,* IIHR Rept. No. 274.
Vita-Finzi, C. (1973) *Recent Earth History,* Macmillan,
    London.
Vries, de M. (1973) River bed variations - aggradation
    and degradation. Inter. Seminar on Hydraulics of
    Alluvial Streams, IAHR, New Delhi, India.
Wyrwoll, K-H (1977) Late Quaternary events in Western
    Australia. *Search,* 32-34.
Wyrwoll, K-H (1984) The sedimentology, stratigraphy and
    palaeoenvironmental significance of the Late Pleistocene
    alluvial fill: the central coastal areas of Western
    Australia. *Catena,* 11, 201-218.
Wyrwoll, K-H (1988) Determining the causes of Pleistocene
    stream-aggradation in the central coastal areas of
    Western Australia. *Catena,* 15, 39-51.
Wyrwoll, K-H and Dortch, C.E. (1978) Stone artifacts and
    an associated diprotodontid mandible from the Greenough
    River, Western Australia. *Search,* 9, 411-413.
Waters, R.R. (1985) Late Quaternary alluvial stratigraphy
    of Whitewater Draw, Arizona: Implications for regional
    correlation of fluvial deposits in the American southwest.
    Geology, 18, 705-708.
Waters, R.R. (1986) The geoarcheology of the Whitewater
    Draw, Arizona. *Anthropol. Papers Univ. of Arizona,*
    No. 45, Univ. Arizona Press, Tuscon.
Young, R.W., Nanson, G.C. and Bryant, E.A. (1986) Alluvial
    chronology for coastal New South Wales: climatic
    control or random erosional events? *Search,* 17, 270-272.

8
Stratigraphy, Sedimentology and Late-Quaternary Chronology of the Channel Country of Western Queensland

Gerald C. Nanson
Robert W. Young
David M. Price

Department of Geography
University of Wollongong
Wollongong, N.S.W.

Brian R. Rust

Department of Geology
University of Ottawa
Ottawa, Canada

I.   INTRODUCTION

The Channel Country, part of the 1.14 mill km^2 Lake Eyre Basin (Fig. 1), is a vast and complex array of fluvial landforms along the inland river systems of western Queensland. Drainage is centripetal to Lake Eyre, located 15 m below sea level in the southern part of the Simpson Desert of South Australia. More than 90% of the area is less than 500 m in elevation, having a surface that is flat to undulating with extensive sand dunes, exposed bedrock and gibber plains. River gradients are generally less than 0.0002 and drainage and channel patterns are a complex function of low slopes, highly variable flows, muddy sediment loads, broad shallow valleys, encroaching dune fields and probably local crustal warping. For the most part channel patterns are braided over the surface of very extensive floodplains formed of mud. Narrow anastomosing

FLUVIAL GEOMORPHOLOGY OF AUSTRALIA
ISBN   0 12 735660 6

channels dissect this surface, interconnecting wider and
deeper permanent water holes, some of which appear like the
disconnected remnants of a much larger channel system within
the floodplain. Apart from sand in some of the anastomosing
channels and water holes, almost all the Channel Country
floodplains are surfaced with dark self-mulching clay soils,
deeply cracked and sometimes with gilgai patterning
(Dawson and Ahern, 1974). An alluvial sand body lies several
metres beneath the floodplain surface. Aeolian dunes
locally intermingle and interact with the floodplain
alluvium.

The objective of this paper is to review what is already
known about Channel Country floodplains, to present new data
on their alluvial and aeolian stratigraphy and chronology,
and to examine the origin of pelleted muds that dominate
the present sediment regime.

II. HYDROLOGY

Rainfall in the Lake Eyre Basin varies from 400-500 mm y^{-1}
in the headwaters to less than 100 mm y^{-1} in the Simpson
desert. Being arid and of such low relief, basin hydrology is
controlled by occasional incursions of moist tropical air
from the north. These spill over the Barkly, Selwyn and
Great Dividing Ranges (Fig. 1) during the northern
Australian monsoon (November-March) but this source of
precipitation is erratic, with rainfall variability being
almost the highest in Australia (Kotwicki, 1986).
Consequently, the rivers do not flood every year and Lake
Eyre fills only a few times each century. Flow occurs most
years at Currarva near Windorah on the Cooper (mean annual
flow of 3.35 x 10^9 m^3) but much is lost by seepage and
evaporation in the myriad of distributary channels,
floodbasins, clay pans, temporary lakes and aeolian dunes,
before the floodwaves reach Nappa Merrie near Innamincka
(2.06 x 10^9 m^3) (Kotwicki, 1986). Downstream, where the
Cooper passes through the dune fields of the Simpson Desert,
these losses are even greater. Only 7.2 x 10^8 m^3 of mean
annual flow enters Lake Eyre, although in most years there
is no input. There is a similar hydrological pattern in
the Diamantina-Georgina sub-basin (Kotwicki, 1986).

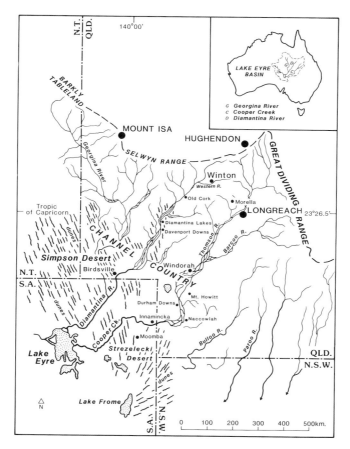

FIGURE 1. *The northeastern part of the Eyre Basin showing the locations of study sites on the Diamantina, Western and Thomson Rivers and Cooper Creek.*

III. PREVIOUS RESEARCH

The first detailed reconnaissance of the Channel Country rivers was conducted by Whitehouse (1948), who argued that many of the present channels and other alluvial features were essentially relict from Pleistocene flow regimes, a position adopted by several subsequent workers (Rundle, 1976; Veevers and Rundle, 1979; Rust, 1981). Whitehouse believed that the ubiquitous floodplain muds "were formed by the ponding of the streams in the past geological age when rainfall was tremendously heavy", and that "some of these ponds then were comparable in area with Lake Eyre of today". Rundle, Veevers and Rust argued that the braided channels were relict because, from the air, it appears that the anastomosing channels bisect and are therefore more recent than the braids. Furthermore, braids had never before been described as forming in such fine-grained mud; they are a pattern usually characteristic of abundant sand or gravel transport. These authors proposed that the braided pattern was the result of a mud drape overlying an extensive, braided sand sheet formed under some prior flow regime, with the anastomosing channels representative of the present mud-dominated regime.

Interestingly, Mabbutt (1967) never subscribed to the relict theory, noting instead that the braided channels are characteristically superimposed across anastomosing channels, "as though indicating a more direct course taken by floodwaters at discharges above bankfull". He also observed that braided channels are more common, and anastomosing less so, where the floodplain is widest.

In a detailed study of Cooper Creek in the vicinity of Naccowlah and Durham Downs, Nanson *et al* (1986) showed that the braided and anastomosing channels make up two contemporary coexistent networks in what today is a mud-dominated river system. During floods, water can extend over the entire floodplain (more than 60 km wide south of Windorah) and both the braided and anastomosing channels are active, whereas at lower stages only the anastomosing channels carry water. With abundant ultra-fine clays being transported over very low gradients, braiding is possible only because the muds are moved as a bedload of sand-sized aggregates with somewhat lower density than quartz sand (Nanson *et al*, 1986). The braided pattern is not a mud drape replicating an underlying sand sheet of braid bars; drilling and excavation revealed that surface channels were quite independent of subsurface palaeochannels in the sand sheet. Furthermore, the underlying sand was shown by Rust and Nanson (1986) to have been deposited as point bars by an array of well-defined meandering channels, larger and more laterally active than any in the present Cooper system. The faint outline of these subsurface palaeochannels and migratory scroll bars can be recognized through the surficial floodplain muds near Mount Howitt (Fig. 1) (Rust and Nanson, 1986, Fig. 7) but generally the mud thickness is now so great that it obscures the meanders.

The chronology of such a marked change in the flow regime of these rivers is important for understanding climatic and flow-regime changes in low latitudes, about

which there is still relatively little known. Whitehouse
(1948, p.25) proposed that most of the fluvial landforms
of the Channel Country were the product of "the period of
the Great Ice Age in high latitudes. In middle and low
latitudes all over the earth at that time it was period of
tremendous rainfall...".

As part of his study of the Quaternary history of the
Strzelecki and Simpson dune fields, Wasson (1983a, 1983b)
showed by dating charcoal fragments that early meandering
anabranches of the southern part of Cooper Creek (Gidgealpa
Palaeochannels north of Moomba) were laterally active some
time earlier than 12 ka BP. At nearby Dog Bite Lake a
minimum age for sandy fluvial deposition was c. 22 ka BP.
Callen *et al.* (1983) have dated pedogenic carbonate from
the same area, obtaining ages of 22 ka - 35 ka BP in
fluvial sandy deposits underlying dunes and fluvial muds.

The stratigraphic sections described by Nanson *et al.*
(1986) and Rust and Nanson (1986) for the Cooper contained
no organic carbon or charcoal suitable for dating. However
a selection of thermoluminescence (TL) dates showed the
overlying mud unit to range from modern at the surface to
50 ka - 80 ka BP just above its contact (at 2.5-7.0 m) with
an extensive underlying sand sheet. The upper sand strata
dated by TL at 200-250 ka BP. From this it appears that
the muds are modern at the surface but that the flow regime
of the Cooper changed from a mixed load but dominantly
sand-transporting system to a mud regime somewhere between
80 ka to 200 ka BP.

The most detailed evidence for aeolian activity is for
the central and southern part of the Cooper Basin (Wasson,
1983a, 1983b, 1986; Callen *et al*, 1983; Gardner *et al*,
1987). These studies show two phases of Holocene dune
activity; one at about 2-3 ka BP and a presently-mobile
active phase. Pleistocene aeolian deposits dated in the
range 13-23 ka BP with a lower unit yielding TL dates of c.
240 ka BP. All four units (two Holocene and two
Pleistocene) contain a mixture of quartz sand and sand-sized
clay aggregates of similar composition to those from the
alluvium of Cooper Creek. Indeed, it is possible that the
structure and origin of aeolian and alluvial aggregates
are the same but this has not been demonstrated. Until now,
the chronology of dunes in the northern part of the basin
has not been investigated, yet they probably form part of
the sediment source of these rivers, possibly even
influencing the process of alluviation.

IV. ALLUVIAL STRATIGRAPHY AND CHRONOLOGY

Major changes in the alluvial stratigraphy of the arid
Lake Eyre Basin are determined largely by climatic and flow
regime changes in the headwater catchments. For that
reason it was decided to focus new work on the accessible
upper reaches of the Cooper and Diamantina Rivers.
Stratigraphic sections were augered and excavated across
the floodplains of the Thomson River at Longreach (Fig.
2) and the Western River at Winton (Fig. 3). The Diamantina
floodplain was augered at Old Cork Crossing, Brighton Downs
and Davenport Downs (Fig. 4). These observations are
compared with those obtained by Rust and Nanson (1986) and
with new information presented here for the central Cooper
between Windorah and Naccowlah (Figs. 1 and 5).

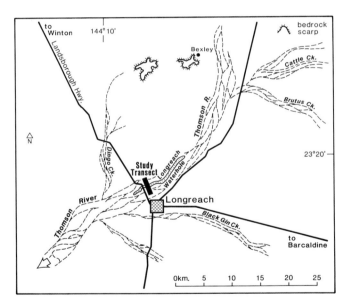

*FIGURE 2.   The study transect on the Thomson River at
Longreach.*

A. *Windorah-Naccowlah*

Here the braid channels are 20-50 m wide, 0.5-0.8 m deep
and are separated by broad lozenge-shaped bars up to 500
m wide. Anastomosing channels are commonly only 20-30 m
wide and 2-4 m deep, although permanent waterholes are about
10 m deep, 60-100 m wide and several kilometres in length.

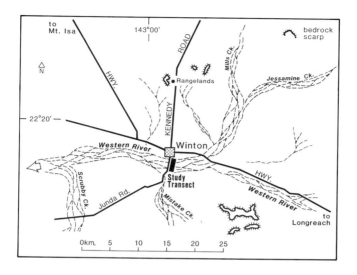

FIGURE 3. The study transect on the Western River at Winton.

A series of holes was augered to between 2 and 11 m depth in anastomosing channels, point bars, levees, braid channels, braid bars, splays and sediment sinks, the latter being backswamp areas remote from major channels, where flood-water ponds and reticulate drainage patterns texture the surface (Figs. 5 and 6). In general, the stratigraphy at Naccowlah shows a 2 to 9 m thick mud sheet covering a unit of relatively clean medium to coarse sand (Fig. 7). The sand unit may be a largely continuous sheet or alternatively it could be formed of a series of partially coalescing sand tongues or ribbons imbedded in overbank floodplain mud. Only further detailed drilling will answer this point. However, because of its almost continuous nature, it is referred to here as a sand sheet. The mud is structureless, massive but cracked into small and large blocks near the surface. The presence of slickenside surfaces indicates that these upper layers have been thoroughly stirred by a self-mulching process associated with wetting and drying. Mud, sand, organic matter and charcoal fragments drop as much as a metre down cracks during dry periods (the charcoal fragments date as modern), and occasional thin laminae and vertical stringers of fine sand are present within this zone. Beneath this uppermost mixed zone the mud is dry, hard, homogeneous, compacted and with few fissures, taking on a polished (marble-like) appearance when excavated.

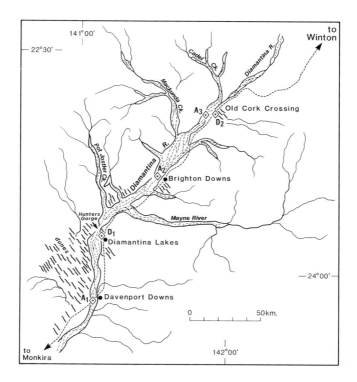

FIGURE 4.   Study sites on the Diamantina River.   D represents dune sites and A alluvial sites.

These lower layers were presumably also well mixed when they were nearer the surface.

The boundary between the mud and underlying sand is usually clearly defined, although lower muds may be sandy and the upper sand strata usually consist of fine sand and mud with mud balls and intraclasts, typical of certain point bar deposits.   Beneath this, the sand is mostly clean and dry, however, mud laminae and intraclasts are also present in places.   Calcite and gypsum indurate the lower sandy muds and muddy sands near the contact, with well defined rhizomorphs and nodules of these precipitates present in some locations.

Rust and Nanson (1986) described abundant sedimentary structures in the form of ripples, dunes and point bar sequences indicative of deposition in sinuous, laterally migrating channels.   The gross palaeochannel morphologies detected by augering support this, revealing meandering

FIGURE 5. (a) Oblique aerial view of the Cooper looking north towards Durham Downs showing sinuous anastomosing anabranches and the braided floodplain surface. (b) Oblique aerial view of braided muds on the floodplain surface and a low sinuosity anabranching channel of the anastomosing system (Bogala Waterhole) 45 km south of Durham Downs.

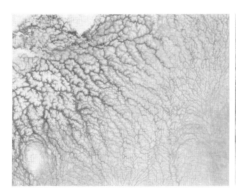

FIGURE 6. (a) Vertical air photo of a floodplain backswamp (sediment sump) 20 km south of Naccowlah. Horizontal field of view, 3.4 km. (b) Oblique aerial view of a sediment splay from the downstream end of a branch of Pritchella Waterhole, 6 km east of Durham Downs. An abandoned splay is visible between the two waterholes.

thalwegs and channel infills. In detail, the flow structures near the top of the sand sheet consist of small-scale trough cross-stratification, planar sets of alternating sand and mud and tabular cross-beds of fine sand interbedded with mud and containing mud intraclasts. With increasing depth the sand becomes cleaner and coarser, containing

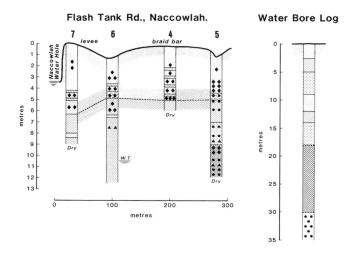

FIGURE 7. A stratigraphic section of the levee and braided floodplain immediately east of Naccowlah Waterhole and a water bore log from the same vicinity. Key is in Fig. 8.

granules and a few pebbles; large-scale tabular cross-beds prevail over less common large-scale trough cross-beds. These structures occur in sets that dip up to 14°, and indicate that the palaeoflow was parallel to the strike of these sets (Rust and Nanson, 1986, Fig. 6b). In other words, the dipping strata were formed by lateral accretion of point bars as an upward fining sequence from mostly large tabular cross-sets of medium to coarse sand and granules representing the base of a channel, to small-scale bedforms and mud laid down at the bar crest or as over-bank deposits.

    1. *Underlying Stratigraphy.* Drilling of a water bore to 35 m in the floodplain was observed near Naccowlah Water-hole (Fig. 7). Down to about 20 m there are two repeating sequences of mud overlying sand. Beneath 20 m the fluvial sands become progressively coarser until gravel and saline water were encountered at the base. These coarse sediments may be from a tributary stream as the drill site was only a few km from the valley side. Drill logs presented by Rust (1981) and Veevers and Rundle (1979) elsewhere in the Channel Country usually show cycles of sand and mud in the upper 100 m of alluvium. They also show a tendency for a greater total proportion of mud in the Cooper alluvium downstream of Innamincka, contrasting with a greater

proportion of sand in the cores nearer Windorah. This is probably related to the greater frequency of palaeofloods and the greater transport power of these floods in the upper Cooper compared to their frequency and effect when they encounter lower gradients in the Simpson/Strzelecki Desert below Innamincka.

2. *Levee Bank and Floodplain Stratigraphy and Chronology.* Augering of levee banks along the Naccowlah and Tabbarrah (near Durham Downs) Waterholes shows a mud/sand contact somewhat below that of the general floodplain (Fig. 7). This suggests that the same waterholes are the remnants of broad sandy channels that contracted during the shift from a sand- to mud-dominated sediment regime, a conclusion supported by stratigraphic and chronological data from Longreach (below). However, it is also very likely that some waterholes are formed by deep scouring of anastomosing channels into the sand body at depth (Nanson *et al*, 1986; Rust and Nanson, 1986), for they are often found at points of flow constriction caused by sand dunes or bedrock.

The topography of the shallow floodplain channels and the surface morphology of the underlying sand sheet exhibit little correspondence (Fig. 8), confirming that the contemporary braided channel pattern bears no relation to the position of channels that transported the prior sand load. The sand unit at Excavation 1 (Fig. 8) TL dates at c. 200-250 ka BP whereas the mud overburden dates at c. 50 ka BP (Rust and Nanson, 1986), results which accord reasonably well with TL dates presented below for the headwaters of the Cooper and for the Diamantina. Particularly significant is a correlation obtained between a sediment TL date and a U/Th date from the same site. The sand sheet at 6 m in Excavation 4 (Rust and Nanson, 1986) TL dated at 232 ± 3.5 ka BP whereas the calcite from a large calcite and gypsum rhizomorph within the sand yielded a U/Th date of 245 ± 74-60 ka BP. Clearly, the rhizomorph must be younger than the surrounding sediment, however, the error brackets on the U/Th date allow for this relatively small discrepancy. What is of particular interest is that the carbonate is apparently nearly as old as the host sediment, something previously not suspected for desert carbonates that in Australia have been evaluated mostly by conventional radiocarbon dating. A radiocarbon date of 35 ± 4 ka BP for this rhizomorph is an illustration of the magnitude of the error that can result from radiocarbon dating a soil carbonate to obtain a minimum depositional age for a geomorphic unit.

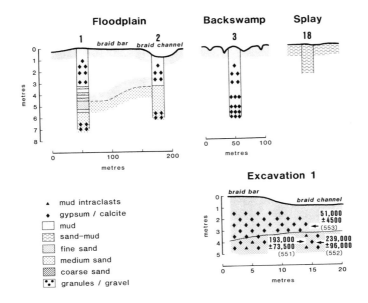

FIGURE 8. Stratigraphic sections of a braid bar and channel on the floodplain, a backswamp (shown in Fig. 6a), a splay (shown in Fig. 6b) and the TL dating site at Excavation 1 near Naccowlah Waterhole.

Interestingly, gypsum within the rhizomorph dated at 61.4 + 16.4 - 15.2 ka BP, suggesting considerable separation of time between calcite and gypsum deposition. The former appears to date about the time the sand was deposited, or soon after, and indicates a semi-arid climate, whereas the gypsum dates from the early phase of mud deposition and is indicative of extreme aridity (Duchaufour, 1982).

3. *Backswamps.* Away from anastomosing and braided channels are backswamps (also called backplains or sumps) characterized by a reticulate drainage pattern (Fig. 6a). They are located where flood waters pond and mud is preferentially deposited. Whitehouse (1948) regarded them as ancient features and Mabbutt (1967) as contemporary, the latter describing the surface self-mulching cracking clay soils as gilgai. Augering in Wilsons Swamp (at the conflucence of the Cooper and Wilson Rivers) showed uniform mud to 6 m (Fig. 8). A stock-water dam in a swamp near Durham Downs was excavated down about 8 m without striking sand (J. McDonald, 1985, pers. comm.). These areas

appear to be sediment sumps created by minor depressions
in the floodplain topography and are probably still forming.
They flood for only a few weeks every year or so and are
dry for the rest of the time.

4. *Splays.* A shallow auger hole in a splay at the downstream end of Pritchella Waterhole (Figs. 6 and 8) revealed
sandy mud, the sand probably being scoured from the
adjacent waterhole during floods. The great spatial extent
of muddy alluvium, compared to the limited extent of
channels and waterholes deep enough to intersect the sand
sheet, means that relatively little of the underlying sand
is being reworked.

5. *Channel Bends.* Anastomosing channels intersect the
top of the sand sheet which is commonly indurated with
carbonate; consequently they usually do not incise into the
sand to any appreciable depth. Figure 9 shows a thin unit
of contemporary channel sand sitting directly on the sand
sheet beneath a section through a channel bend. The thick
mud unit adjacent to the point bar is similar to that at
the cutbank and confirms that these channels are not
migratory, for there is no laterally accreting sand unit.

B. *Thomson River at Longreach*

This Cooper tributary has an almost entirely braided
pattern of shallow channels over the floodplain surface,
with Longreach Waterhole the only large channel and
permanent water body in the vicinity. The floodplain is much
narrower than downstream and exhibits fewer landform types.
The surface soils and braided floodplain channels are
similar to those elsewhere in the Cooper Basin, as is
Longreach Waterhole itself. Perhaps the most distinctive
difference is that at Longreach there is a low but
recognisable terrace relatively close to the main channel,
something not observed on the lower Cooper or Diamantina.
To compare the stratigraphy of this site with the other
locations studied, a number of holes were augered and 4-7
m deep trenches excavated along a transect perpendicular
to the river (Figs. 2 and 10). This is remarkably similar
to Naccowlah, 550 km downstream. A`capping of about
2 m of mud overlies a sand sheet of clean medium to coarse
sand and a few pebbles and granules, exhibiting tabular
and trough cross-stratification. However in contrast to
Naccowlah, there was no evidence that these palaeochannels

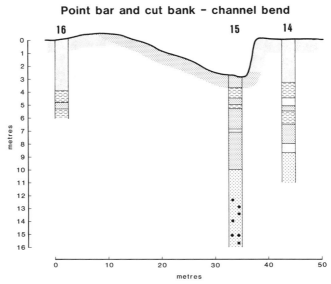

FIGURE 9.   Stratigraphic section through a channel bend
east of Durham Downs.   Thick mud units on both sides of the
sandy channel floor indicate no channel migration (Key as
in Fig. 8).

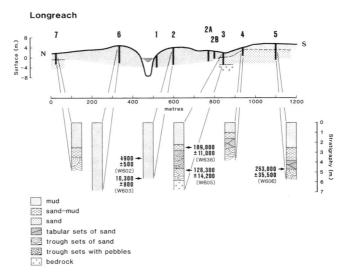

FIGURE 10.   Stratigraphy of the Thomson River near
Longreach (Figs. 1 and 2).

were sinuous; it is probable that they were wide, shallow and braided, characteristic of a narrower valley and steeper gradient. Between the mud and the sand sheet is a unit of sandy mud which may well be the overbank deposit associated with the sand-dominant phase, a conclusion supported by TL dating. Adjacent to Longreach Waterhole the mud unit is very thick (beyond the reach of the excavator). At the conclusion of the sand-transport phase, it appears that a wide sandy channel contracted and partially infilled with mud to form the present 6 m deep waterhole (Fig. 10).

Topographic surveying revealed a slightly elevated terrace to the southeast (Fig. 10). The terrace sand dates at 263 ± 35.5 ka BP, similar to the extensive sand bodies in the Naccowlah area. Closer to the waterhole is a buried terrace with sands of 128 ± 14.2 ka BP and overlying sand-mud perhaps slightly younger (Fig. 10). The mud infill and levee adjacent to the waterhole are very much younger; a TL date from nearly 6 m depth was only 10.3 ± 0.8 ka BP. There appears to have been a considerable reduction in the cross section of this waterhole during the Holocene.

C.   Western River at Winton

A similar cross section to that at Longreach was excavated at Winton on the Western River in the upper Diamantina (Figs. 3 and 11). There is no distinct waterhole, but rather a series of usually dry anastomosing channels separated by well defined muddy islands. Excavations were limited to 4 m in depth due to the small size of the excavator, but from the base of the channels, this meant 6-7m below the floodplain surface.

The stratigraphy and alluvial chronology are somewhat different to those of the Cooper system. The upper muddy unit seems physically identical but thicker. As at Longreach, and in contrast to Naccowlah, it is relatively young (<20 ka BP), except 6 m below the surface at the north end of the section, where an age of 85 ± 10 ka BP was obtained. Beneath the uniform mud unit lies a zone of interbedded sand and mud which gave ages of 41.5 ± 3.4 ka BP and 35.8 ± 2.7 ka BP and may represent a substantial period of alternating mud- and sand-load transport not detected in the Cooper system.

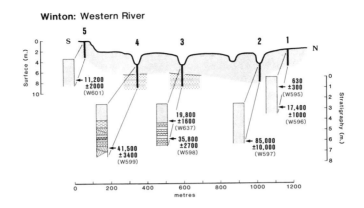

FIGURE 11. Stratigraphy of the Western River near Winton (Figs. 1 and 3) (Key as in Fig. 10).

D.   *Diamantina at Davenport Downs and Brighton Downs*

Two auger holes through the bed of anastomosing channels (Fig. 12) support the observations from further upstream on the Diamantina system at Winton. Firstly, the upper mud unit is again thicker; secondly, it tends to be younger, and thirdly, there is again a zone of interbedded fine sand and mud beneath the mud unit. The section at Davenport Downs (Fig. 12a) shows a channel cut in an upper mud unit. Below this is a unit of interbedded muddy sands and a substantial mud unit beneath that again, the latter equivalent in age to the deep mud unit at Winton (about 85 ka BP). It is thought that an older sand unit lies beneath the limit of augering but this could not be verified here. However, at Brighton Downs this sand body was intersected and it yielded a TL age of 97 ± 10 ka BP, very close to that of the buried terrace at Longreach (Fig. 10).

The Diamantina mud unit can be a uniform dense mud with a basal age of between 85 and nearly 100 ka BP (as at the north end of the Winton transect and at Brighton Downs), or there can be interbedded muddy sands within a part of this column. The younger of the two Cooper sand units would seem to be present at depth, but confirmation of this, and the existence of a 200-260 ka BP sand unit as occurs on the Cooper must await TL dates from deeper excavations.

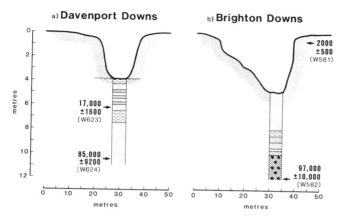

FIGURE 12. Stratigraphy of the Diamantina River at Davenport Downs and Brighton Downs (Figs. 1 and 4) (Key as in Fig. 8).

V. THE ALLUVIAL MUD

Because clay in these rivers is transported mainly as sand-sized aggregates, mud braids can occur (Nanson *et al*, 1986). Nevertheless, the origin of the mud aggregates is problematic. Bowler (1973) suggested the desiccation of mud in the presence of saline groundwater is responsible for the formation of clay pellets in aeolian deposits flanking the shores of many dry lakes. This mechanism seems applicable neither to the vast alluvial sheets in the Channel Country (Nanson *et al*, 1986) nor probably to the clay pellets in the dunes in some parts of the adjacent Simpson Desert (Gardner *et al*, 1987). An alternative explanation is that they derive from weathering of mudstones and labile sandstones which outcrop over extensive areas of the basin, and that their pelletised form is maintained by frequent wetting and drying of swelling clays. Rain showers or flooding, followed by intense baking under a hot sub-tropical sun, appear to be a sufficient mechanism. This explanation is supported by observations of soil profiles on Cretaceous outcrops well above modern floodplains at Morella quarry (Fig. 1) and at Bexley Hill near Longreach (Fig. 2), and from initial laboratory experiments (below).

The fluffy "pop corn" structures, so typical of the desiccated surface of the alluvial mudsheets, are also very well developed in deep chernozems at Morella and in lithosols on the steep flanks of Bexley Hill. Moreover, thin

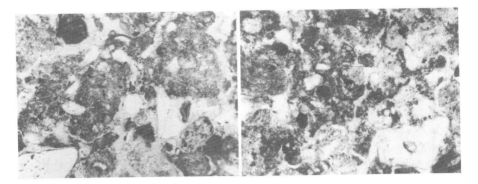

FIGURE 13. Thin sections showing clay pellets and quartz grains in a chernozem profile developed on a weathered regolith over Mesozoic mudstone at Morella quarry. Horizontal field of view, 0.9 mm.

sections of these soils showed well-developed clay aggregates similar to those in the modern alluvium (Fig. 13). Thus it seems that the pellets are formed primarily by pedogenesis of the weathering products of these rocks. Thence they are transported downslope to the surrounding mudsheets where there would be a long period of churning in vertisol profiles. Seasonal desiccation and reworking of the alluvial muds appear to play a major role, for pellets are best developed in the upper layers of the alluvium. With burial, they become compressed and deformed; under 2-3 m of overburden all interstitial spaces are lost and they inter-lock in a tight-fitting three dimensional pattern (Fig. 14). Laboratory experiments show that the surface pellets are stable throughout several hours of transport as bedload in a hydraulic flume (Nanson *et al*, 1986) and initial results show that they can be disaggregated and reaggregated in the laboratory by numerous cycles of wetting and drying under a sunlamp (Maroulis, in prep.).

X-ray diffraction shows that, in the upper part of the basin, hillside and alluvial clays alike consist mainly of somewhat degraded mixed-layer smectites, with minor kaolinite and quartz, and occasional hydrobiotite. No special circumstances need therefore be invoked to explain the development of clay pellets here; they are the product of the swelling and contraction of expansive clays under the prevailing subtropical, seasonally-humid, semi-arid (Koppen's B.Sh) climate.

The relationship between bedrock and alluvial properties is more complex in the central and southern parts of the basin where Mesozoic sediments are extensively mantled

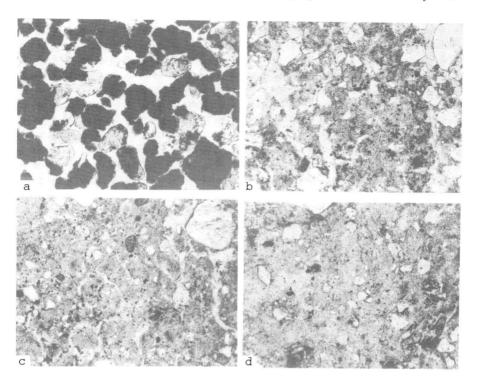

FIGURE 14.   Alluvial muds from near Naccowlah in
various stages of compaction as a result of floodplain
accretion. (a) Loose surface mud showing pellets (black)
and quartz sand grains (translucent). (b) Thin section from
20 cm beneath the floodplain surface.   (c) Thin section at
1.0 m depth.   (d) Thin section from 2.0 m depth.   Note how,
with depth, the mud compacts to an almost structureless
matrix around the quartz grains.   Horizontal field of view,
1.6 mm for (a) and 1.2 mm for the remainder.

with Tertiary sediments and deep-weathered profiles.   Many
of the relict profiles yield much fine-grained sediment high
in kaolinite; samples from Morny Station west of Windorah
consist of 26% silt and 72% clay. Here is the reason for
the apparent downstream increase of kaolinitic clay in the
alluvium (cf Nanson *et al*, 1986).   Nonethless, smectites
are still present in sufficient quantities to cause
shrinking and swelling that will form pellets.

VI.   AEOLIAN DUNE ACTIVITY

Aeolian dunes flank the Diamantina and Cooper for much
of their lower reaches, invading the floodplain in many
places. Some of the most northerly occur in relatively small
dune fields between Davenport Downs and Old Cork Crossing
on the Diamantina (Fig. 4). To test whether these dunes
were active at times corresponding or contrasting with
episodes of fluvial deposition, TL samples were collected
from a single large dune adjacent to the homestead at
Diamantina Lakes, and from a low roadside dune on the east
bank of the Diamantina River at Old Cork Crossing (Fig. 15).
The dune at Diamantina Lakes consists of several sand units
separated by clayey sands. The uppermost unit is 4 m of red
sand. Beneath it is 1 m of reddish clayey sand which gave
a TL date of 32.5 ± 6 ka BP. A further 1 m of red sand is
followed by 1.5-2.0 m of brownish clayey sand which dated at
274 ± 22 ka BP. It possibly represents the same phase of
dune building as that dated at c.220-240 ka BP in the
Simpson Desert by Gardner *et al.* (1987), or it could be even
earlier. At Old Cork the low dune consisted of 3 m of red
sand over 1-2 m of red clayey sand (Fig. 15). The lower unit
of this dune, which has weathering features like those in
the younger dated unit at Diamantina Lakes, gave a TL date
55.5 ± 6 ka BP. These reconnaissance dates indicate that
the dune chronology roughly spans that of the adjacent
alluvium. Furthermore, they suggest that there has been
aeolian activity during both the sand-and mud-dominated
fluvial phases, although there is clearly a need for further
detailed information.

VII.   DISCUSSION AND CONCLUSIONS

Interpretations of landscape evolution and contemporary
fluvial dynamics for the Channel Country have changed
considerably since the 1940s. While detailed research has
been sparse, it is now apparent that there is great potential
here for providing environmental data on Quaternary climatic
and hydrological changes in central Australia. The time
span goes well beyond the range of radiocarbon dating and,.
furthermore, such arid environments are poor sites for
organic carbon deposition and preservation (Callen *et al,*
1983). Thermoluminescence analyses have permitted
chronological interpretations within the last 300 ka.

Diamantina Lakes

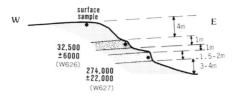

Old Cork Crossing

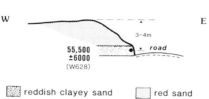

reddish clayey sand   red sand
brownish clayey sand   • TL sample

FIGURE 15. Dune stratigraphy along the Diamantina River. Sections not to scale. Figure 4 shows locations.

Correlation of TL results with a U/Th date, and with radio-carbon ages for alluvial deposits elsewhere (Nanson and Young, 1986; Gardner *et al*, 1987), gives confidence in these interpretations.

Deep drill holes show evidence of alternations between mud- and sand-dominated sediment regimes, with the present clearly mud-dominated. The sandy phases have been characterised by meandering, laterally-migrating channels in the middle reaches of the Cooper. However, headwater palaeochannels appear to have been sandy and braided. Deep bore holes show the stratigraphy to consist of alternating units of alluvial sand and mud. Thermoluminescence dates presented here and elsewhere (Rust and Nanson, 1986), show the uppermost sand sheet consists of two sand bodies, one dating at about 100-120 ka and another at about 200-250 ka. The younger was not detected on the middle reaches of the Cooper but may exist there at locations close to the present water holes. The overlying mud unit dates from about 85 ka to the present and probably represents a shift towards greater climatic aridity, a conclusion supported by the U/Th dating of gypsum. Furthermore, there is good evidence, particularly at Longreach, that the original sand-load channels contracted considerably in size with the onset of the present mud-dominated regime. A number of waterholes in the Channel Country could be the remnants of what were large sand-load channels operative about 100 ka. On the

Cooper and lower Diamantina, these waterholes still retain
what was possibly the original sinuosity of these sand
channels (Fig. 6b) (Figs. 2 and 6 in Nanson *et al*, 1986).
However, many waterholes appear to be more recent scour
features, for they are of lower sinuosity than the palaeochannels (Rust and Nanson, 1986).

    The alternating sand and mud units at depth probably
represent alternating "humid" and "arid" climatic shifts.
However, these may relate to thresholds of change rather
than major shifts in the hydrological regime. The dune
chronology of the Diamantina suggests that both the mud and
sand phases were fairly arid. A change in the frequency
of extreme storms producing runoff from the highlands might
have been sufficient to engender shifts in flow and sediment
regime evident in the stratigraphy.

    One notable characteristic of these rivers is their
ability to switch their dominant transport load from sand
to mud. This suggests a commensurate change in the supply
of sediment. However, an alternative explanation is that
both sediment types are present all the time but that there
is shift in the river's ability to transport and deposit
its mixed load. The sand unit is characterised by numerous
mud intraclasts, laminae and drapes. Undoubtedly these
dominantly sand-load rivers were also transporting a
considerable fine suspended load. It may well be that the
mud unit that presently covers almost all of the Channel
Country has resulted not from a wholesale shift in the catchments' source of sediment, but from the reworking and
redistribution of already abundant overbank sediment. In
other words, the muddy phases may represent periods when
the rivers had insufficient energy to move the abundant sand
and mud-pellet load through the existing large channels.
As a result, many of these channels "silted up" and the
displaced overbank flow has reworked the pelleted floodplain muds into the present network of braid bars and
channels. Locally flow is concentrated on this surface and
cuts anastomosing channels into the compacted muds,
although these channels still interact hydraulically with
the surficial braided pattern during times of flood
(Nanson *et al*, 1986). At other locations remnants of the
ancient sand-load channels are preserved as waterholes.
These are fairly straight in the upper reaches where the
palaeochannels were probably braided, but some are clearly
sinuous in the middle and lower reaches. In places where
flow energy is concentrated, recent waterholes have probably
been incised into the floodplain muds. At the onset of more
humid hydrological regimes, it is probable that the waterholes are the first channels to become erosionally active.

They are large and able to concentrate considerable erosive energy and would expand or start to laterally migrate, reworking floodplain sediments as they did so.

The source of the abundant mud load of these streams has never been thoroughly addressed before. Data presented here show that no special conditions need to be invoked to explain their extent and characteristics. They have been derived from the vast outcrops of Mesozoic mudstones and labile sandstones and from extensive deep weathering profiles. Pedogenesis under an arid to semi-arid climate accounts for their highly pelleted structure.

Detailed research in the inland plains of southeastern Australia has demonstrated clearly that climatic changes in the last 40,000 years or so have led to major shifts in the hydraulic regime of the Murray-Darling system (Bowler, 1978, 1986). Yet the chronology now emerging for the Channel Country indicates that last glacial and Holocene climatic changes were of little consequence; it seems that major shifts in hydrologic regime long predate them. Sand-dominated phases appear to have been operative about 240 ka and 120 ka BP, and carbonate precipitation at that time suggests available moisture but in a semi-arid climatic regime. Mud deposition dates from about 85 ka BP and is probably indicative of the considerable increase in aridity as evidenced by dated gypsum. The sand-dominated phases appear to coincide with major world-wide interglacials at about 100-120 ka and 200-250 ka.

To conclude, it is interesting to speculate why the present interglacial climate is not associated with a sand-load regime in the Channel Country rivers. It appears that the present interglacial is so far not as humid in the Eyre Basin as were previous interglacials, a conclusion that must await further independent evidence.

ACKNOWLEDGEMENTS

The authors are particularly grateful to Stephen Short (A.N.S.T.O.) for undertaking the U/Th dates, to Graham Taylor (Canberra C.A.E.) for assistance in the field and to Delhi Petroleum for logistical support and bore log data. This research was funded with research grants from A.R.G.S and the University of Wollongong.

REFERENCES

Bowler, J.M. (1973) Clay Dunes: Their occurrence, formation and environmental significance. *Earth Sci. Rev.*, 9, 315-338.

Bowler, J.M. (1978) Quaternary climate and tectonics in the evolution of the Riverine Plain, Southeastern Australia. In J.L. Davies and M.A.J. Williams (eds.) *Landform Evolution in Australasia*, ANU Press, Canberra, 70-112.

Bowler, J.M. (1986) Quaternary landform evolution. In D.N. Jeans (ed.) *Australia, a Geography, Vol. 1 The Natural Environment*, Sydney Univ. Press, 117-147.

Callen, R.A., Wasson, R.J. and Gillespie, R. (1983) Reliability of radiocarbon dating of pedogenic carbonate in the Australian arid zone. *Sed. Geol.*, 35, 1-14.

Dawson, N.M. and Ahern, C.R. (1974) Soils. Western Arid Region, Land Use Study Part 1. *Qld. Dept. Prim. Ind. Tech. Bull.*, 12, 18-46.

Duchaufour, P. (1982) *Pedology: Pedogenesis and Classification* (Transl. T.R. Paton), Allen and Unwin, London.

Gardner, G.J., Mortlock, A.J., Price, D.M., Readhead, M.L. and Wasson, R.J. (1987) Thermoluminescence and radiocarbon dating of Australian desert dunes. *Aust. J. Earth Sci.*, 34, 343-357.

Kotwicki, V. (1986) *Floods of Lake Eyre*, Eng. and Water Supply Dept, South Australia, Adelaide.

Mabbutt, J.A. (1967) Geomorphology of the area App. 4. In C.M. Gregory, B.R. Senior and M.C. Galloway (eds.) *The Geology of the Connemora, Jundah, Canterbury, Windorah and Adavale 1:250,000 Sheet Areas, Queensland.* Bur. Min. Res. Geol. Geophys. Record 1967/16.

Nanson, G.C., Rust, B.R. and Taylor, G. (1986) Coexistent mud braids and anastomosing channels in an arid zone river: Cooper Creek, Central Australia. *Geology*, 14, 175-178.

Nanson, G.C. and Young, R.W. (1987) Comparison of thermoluminescence and radiocarbon age-determinations from late-Pleistocene alluvial deposits near Sydney, Australia. *Quat. Res.*, 27, 263-269.

Rundle, A.S. (1976) Channel Patterns of Multiple Streams. Unpubl. PhD thesis, Macquarie University.

Rust, B.R. (1981) Sedimentation in an arid-zone anastomosing fluvial system: Cooper Creek, Central Australia. *J. Sed. Petrol.*, 51, 745-755.

Rust, B.R. and Nanson, G.C. (1986) Contemporary and palaeochannel patterns and the Late Quaternary stratigraphy of Cooper Creek, Southwest Queensland, Australia. *Earth Surf. Proc. and Landforms*, 11, 581-590.

Veevers, J.J. and Rundle, A.S. (1979) Channel Country fluvial sands and associated facies of Central-Eastern Australia: modern analogues of Mesozoic desert sands of South America. *Palaeogeogr., Palaeoecol.,* Palaeoclim., 26, 1-16.

Wasson, R.J. (1983a) The Cainozoic history of the Strzelecki and Simpson dunefields (Australia) and the history of the desert dunes. *Zeit. f. Geomorph.,* Supp. 45, 85-115.

Wasson, R.J. (1983b) Dune sediment types, sand colour, sediment provenance and hydrology in the Strzelecki-Simpson dunefield, Australia. In M.E. Brookfield and T.S. Ahlbrandt (eds.) *Eolian Sediments and Processes,* Elsevier, Amsterdam, 165-195.

Wasson, R.J. (1986) Geomorphology and Quaternary history of the Australian continental dunefields. *Geog. Rev. Japan,* 59 (Ser. B), 55-67.

Whitehouse, F.W. (1948) The geology of the Channel Country of southwestern Queensland. *Qld. Bur. Invest. Tech. Bull.* 1, 10-28.

9
History, Palaeochannels and Palaeofloods of the Finke River, Central Australia

G. Pickup
G. Allan

CSIRO
Division of Wildlife and Ecology
Alice Springs, N.T.

V.R. Baker

Department of Geosciences
University of Arizona
Tucson, Arizona

I. INTRODUCTION

Most Australian fluvial geomorphology has been concentrated in N.S.W. coastal rivers. These systems have
short courses, relatively steep gradients and are set in
alluvial plains dominated by fairly recent features which
makes them atypical of most of the continent. The result
is that there is now a danger of developing models of
Australian river behaviour which are untested in and perhaps
irrelevant to the inland area where most of the fluvial
landscapes occur. Studies of inland rivers are relatively few
and virtually nothing exists for arid-zone mountain rivers
which play a major role in setting the pattern of internal
drainage which occurs in the central parts of the continent.
The mountain rivers are of great age and, as such, are
a rich source of information on the forms and processes of
long term erosion which have been obscured elsewhere in the
world by glaciation, orogeny, sea level change or burial
in accumulated sediment. They can also provide information

FLUVIAL GEOMORPHOLOGY OF AUSTRALIA
ISBN 0 12 735660 6

on the extreme events which help determine their morphology
and that of the alluvial systems which lie downstream.   A
growing body of information suggests that fluvial landscapes
in the arid zone are dominated by these large events which
create landforms that survive for long periods.   The history
of mountain rivers cut in bedrock can therefore contain the
key to landscape features in much larger areas.

   This essay summarises recent work in the Finke bedrock
gorge, one of central Australia's larger and most spectacular
river systems.   While the Finke may not be "the oldest river
in the world" as the tourist guides suggest, it has followed
some sections of its course perhaps since the Devonian when
the Alice Springs orogeny occurred (Cook, 1968).   The
associated entrenchment of the river has left a series of
relict features showing a complex erosional history and
some processes of landform development which are still
difficult to explain.   The Finke is also the scene of the
first attempts in Australia to reconstruct flood history
from slackwater deposits and has what is probably the best-
explored slackwater deposit sequence on the continent.   It
therefore offers some insights on the magnitude and frequency
of catastrophic floods and how they are expressed in the
landscape.

II.   GEOLOGICAL SETTING

   The Finke River rises in the Western MacDonnell Ranges,
a set of east-west trending parallel strike ridges and
valleys in rocks of Cambrian and Pre-Cambrian age in central
Australia (Fig. 1).   The Pre-Cambrian rocks consist of the
granites and metamorphics of the Arunta complex which are
flanked by Heavitree Quartzite and Cambrian sedimentaries
(Quinlan and Forman, 1968).   The quartzite is of particular
importance because it is very resistant to both erosion and
weathering and provides the cobbles which make up much of
the Finke bedload for a considerable distance downstream.
These cobbles also occur on some relict surfaces
described below, marking them as former courses of the river.
The river exits the Western MacDonnells by flowing south
through a series of pediments in Brewer Conglomerate and
crossing the Missionary Plain, a broad area of Quaternary
alluvium and sand dunes flanked by laterites.   It then enters
and crosses the Krichauff and James Ranges, a series of east-
west oriented anticline-syncline structures which lie
across the path of the river and post-date it (Fig. 2).

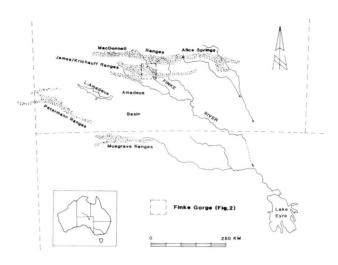

FIGURE 1. Location map.

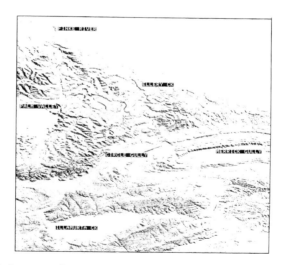

FIGURE 2. Landsat Band 5 image showing the passage of Finke River and Ellery Creek through the Krichauff and James Range anticlines.

Once in the northern part of the Krichauff Range, the river occupies a deep gorge cut in Hermannsburg Sandstone. The initial 12 km of the gorge are straight but after that, it meanders. The meandering pattern is complex and may

reflect antecedence. There are actually two gorges in the
section of river north of the James Range - the one currently
occupied by the river and a second one which is not so deeply
entrenched into bedrock and looks like another meander
train intertwined with that of the main gorge. These
palaeomeanders mainly occur upstream of the Ellery Creek
junction on both the Finke and the Ellery and almost always
contain a fossil bed material of rounded quartzite gravel
and cobbles. The double gorge structure only extends as
far as the anticline which forms the James Range. After
this the river crosses a series of strike valleys and ridges
formed in alternating sandstones and siltstones of Cambrian
to Devonian age. The strike valleys allow the river to widen
out and contain a wide variety of aeolian and fluvial
deposits as well as material weathered *in situ* from the
siltstones. On the south side of the James Range anticline,
the river once again crosses Hermannsburg Sandstone but it
is not deeply entrenched as in the upper part of the gorge.
Instead it occupies a wide valley which contains a substantial volume of recent sediment. Beyond this, the river
leaves the ranges and crosses a series of alluvial plains,
sand plains and dune fields before joining the Palmer River
and draining along the western edge of the Simpson Desert
towards Lake Eyre.

III.   EARLY GEOMORPHIC HISTORY

A.   *Planation Surfaces*

     The Mesozoic and Tertiary history of the central
Australian ranges involved prolonged episodes of deep
weathering and denudation, during which multiple erosion
surfaces evolved (Mabbutt, 1966). During this time the
resistant structures of the ranges were etched into relief
as less resistant rocks and regolith were preferentially
eroded (Fig. 3). Much of this etchplanation probably
occurred after a major phase of Mesozoic lateritisation and
erosive stripping produced spectacular crest bevels.
Unlike the younger, incised landscapes, these ancient summit
surfaces are cut right across dipping bedrock. The higher
crest-bevel surfaces in the Krichauff and James Ranges
probably correlate to the Cretaceous erosion surface that
truncates the Gosses Bluff impact structure of the Missionary
Plain, 80 km northwest of the study area (Milton   *et al,*
1972).

FIGURE 3. Prominent crest bevel surface (left) developed across the northeasterly dipping Hermannsburg Sandstone of the Krichauff Range. A strike valley occupied by Boggy Hole Creek (centre and bottom left) separates the Krichauff and James Ranges (right). Note the deep incision and alluvial fan (left centre) produced by relatively recent incision into the crest bevel surface, which probably formed in Cretaceous to early Tertiary times.

Unfortunately, the lack of accurate topographic mapping for this region precludes a definitive analysis of the time-elevation relationships among various geomorphic surfaces. A number of planation surfaces have been identified (Fig. 4), but the following descriptions and inferred history must be considered hypothetical until future work can be done with accurate elevation data.

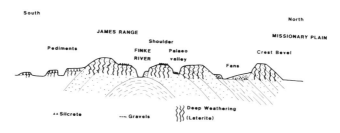

FIGURE 4. Schematic south-north cross section through the Krichauff and James Ranges showing idealised relation- ship of crest bevels to structure. Lag gravels are associated with the shoulder surface (Fig. 5) and the palaeomeander valleys. Pediments flanking the ranges locally display a siliceous cap over lateritic regolith.

Below the level of the high crest bevels, which truncate
various strike ridges (Fig. 3), a prominent erosion surface
is developed on either side of the modern Finke Gorge (Fig.
5). Upstream of Circle Gully, this surface comprises a
discrete shoulder (S in Fig. 5) between the deeply incised
trench of the main gorge and the uppermost elements of the
Krichauff Range. The topography of the shoulder surface
is relatively gentle and rolling, and it is locally littered
with quartzite lag gravels. The hillslopes separating the
shoulder surface from higher remnants of the crest bevel
surface (B in Fig. 5) have rounded, convex upward profiles.
Locally, there are cirque-shaped hollows and remnants of
broad, low-relief valleys which form embayments within the
highest uplands of the Krichauff Range.

The shoulder surface comprises divides between the
modern Finke Gorge segments (M in Fig. 5). It is also
incised by the palaeomeander valley (P in Fig. 5),
described in more detail below. The shoulder surface may
have some structural control induced by the dipping
Hermannsburg Sandstone, but it seems most likely to be the
remnant of a very broad, low-relief valley cut into uplands
which form the ancient crest-bevel surface. The slope
morphology and lag gravels suggest planation during a humid
episode, and the relationship to higher surfaces suggests
a mid-Tertiary age.

FIGURE 5. Shoulder surface (S) developed on
Hermannsburg Sandstone 4 km northwest of junction of Ellery
Creek (E) with Finke River (centre and foreground). Note
rounded slope forms separating shoulder surface from remnants
of crest-bevel surface (B). Palaeomeanders (P) are incised
into shoulder surface but show rounded valley-side slopes.
Modern Finke Gorge (M) shows faceted slopes, especially at
meander scars.

B. Palaeomeanders

The palaeomeander valley begins upstream of the junction of Ellery Creek and the Finke Gorge (Fig. 6). The ancient valley floors occur 15 to 30 m above the modern river level, with an upstream increase in elevation above the modern river, particularly along Ellery Creek. There are also local bedrock benches and embayments within the modern gorge that mark relict portions of palaeochannel floor. A broader remnant of this surface occurs in the Amphitheatre area of Palm Valley (Fig. 7).

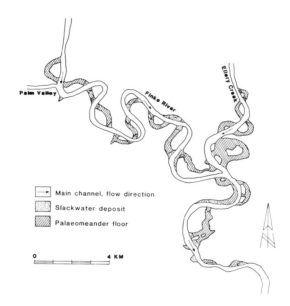

Main channel, flow direction

Slackwater deposit

Palaeomeander floor

0   4 KM

FIGURE 6. Palaeomeander systems of the Finke Gorge.

The palaeomeanders are floored with rounded and highly weathered quartzite gravel overlying bedrock. The gravel is a lag, with other former bed material having weathered to red silt and clay. Locally thick sections of palaeo-meander gravel are cemented by a siliceous matrix into a conglomerate. This shows large cracks and fissures that break across rather than between individual quartzite clasts. The siliceous conglomerate is also associated with ferruginous crusts on the sandstone slopes. These ancient deposits and their relict position in the landscape suggest considerable antiquity (Twidale, 1983), probably at least Miocene age.

FIGURE 7.   *Finke River Gorge downstream of junction with Palm Valley Creek (upper right). Note shoulder surface (S), remnants of crest-bevel surface (B), and palaeomeander valley (P). Amphitheatre (A) may be part of an erosion surface adjusted to the palaeomeander valley level.*

C. *Modern George*

The morphology of the modern gorge is in marked contrast to that of the shoulder surface and the palaeomeanders. The modern gorge sharply cuts across the latter surfaces (Figs. 6 and 7). Slopes are straight, or faceted, especially at meander bends (Fig. 5). Incision to the modern level may have been relatively rapid, in contrast to the palaeomeander incision, because spectacular hanging valleys occur (Fig. 8). These show the rounded slope forms of the shoulder surface.

The indurated crusts associated with the modern gorge are all calcretes. These occur close to the modern river bed and on the lower parts of embayments that correspond to the floors of the palaeomeanders. However, they clearly flank and are considerably younger than the palaeomeander gravels. Near the junction of Circle Gully Creek and the Finke River a particularly thick deposit of calcrete-cemented gravel shows indurated calcic profile morphology with multiple generations of laminae and pisolites. The recemented morphology and case-hard surfaces are termed "Stage VI" when recognised in pedogenic calcretes developed in non-calcareous parent materials under arid and semi-arid climates of the American southwest. The youngest Stage VI calcretes of the southwestern United States are Pliocene in age, while many are Miocene (Machette, 1985).

The modern river gorge of the Finke predates the calcretes possibly making it Pliocene or earlier in age. It formed

FIGURE 8. *Hanging valley in the "Glen of Palms" reach of the Finke River between the junctions with Palm Valley Creek and Ellery Creek. Note the vertical cliff at the cutbank of the modern Finke Gorge and the rounded slopes of the hanging valley developed on the shoulder surface, which forms the skyline.*

during a period of semi-arid to arid climate favouring soil calcification, in marked contrast to the lateritisation coincident with the formation of the palaeomeanders and older surfaces.

D. Drainage Evolution

Drainage patterns during the time of the crest-bevel surface (Cretaceous) cannot be established. However, wind gaps cut into the crest-bevel surface of the MacDonnell Ranges (Mabbutt, 1966) and Gosses Bluff (Milton *et al*, 1972), show that by early Tertiary time streams were cutting across the fold structures of central Australia much as they do today, but with much more subdued topography and rounded valley slopes. By the time of development of the shoulder surface (mid-Tertiary?) there was a broad valley cut through the Krichauff uplands.

There is a highly speculative case that the Finke predecessor occupying the shoulder surface may have drained around the James Range anticline, exiting to the north of Circle Gully and coming out of the Range along what is now the valley occupied by Illamurta Creek (Fig. 2). The Finke headwaters might then have been captured by a stream cutting headward through the James Range anticline (Fig. 9). Alternatively, Illamurta Creek did the capturing of a strike stream graded to the shoulder surface Finke and draining a large part of the western James Range.

FIGURE 9.   Finke River Gorge through northern margin of James Range, north of Circle Gully (C) and Boggy Hole Creek (H).   The ancient Finke, occupying the shoulder surface (S), may have flowed to the left of this picture (west) prior to capture by a tributary creating the prominent water gap at the centre of the picture.

That such captures could have occurred, as hypothesised above, is illustrated by a prominent wind gap south of Circle Gully (Fig. 10).   The gap has rounded side slopes and is cut through the crest-bevel surface.   However, the Finke does not seem to have flowed through Circle Gully, which is filled with weathered sandstone and siltstone but lacks the distinctive quartzite gravels.

The origin of the intertwined palaeomeandering valley and the modern Finke Gorge is probably the most intriguing aspect of the geomorphic history.   It may be that both patterns developed on the shoulder surface, but that the modern gorge was favoured when rapid downcutting occurred. Simultaneous development of portions of the valley is indicated by the palaeomeander immediately south of Ellery Creek junction.   That palaeomeander is filled with younger fluvial sediments than the quartzite lag gravels and may even have conveyed Holocene flood water during the most extreme events.   However, in contrast, the higher, upstream portions of the palaeomeander train are clearly relict and very ancient.

The changes in slope morphologies from the palaeomeanders to the modern gorge parallel observations in the Ooraminna Ranges, 140 km east of the study area, by Twidale and Milnes (1983).   The gentle slopes of Miocene age were converted to near-vertical cliffs in later Cenozoic time.

FIGURE 10. Wind gap (centre) crossing a prominent
crest-bevelled ridge of the James Range from Circle Gully
(right) to a strike valley (left). Finke River (bottom)
crosses the ridge at a water gap just to the right of the
picture.

Regional uplift in the late Tertiary disrupted a Miocene
landscape characterised by local lakes and silicification
(Milnes and Twidale, 1983). The climate changes probably
paralleled those of north-eastern South Australia, where
Callen (1977) recognises subtropical conditions in the
Miocene, followed by semi-arid conditions in the Pliocene,
and aridity in the Pleistocene.

IV. HOLOCENE FLUVIAL HISTORY

    The bedrock geomorphology of the currently active Finke
Gorge was established before the development of the
calcretes. Its Holocene history has therefore been one in
which the major fluvial sedimentary features of the gorge
have developed. This history is essentially a sequence of
great floods separated by periods of minor activity. The
great floods establish the boundary of the channel and the
location of the major in-channel bars. The intervening
periods involve local modification of existing channels by
minor flows but usually within constraints set by large
flood features and the form of the bedrock gorge.
    The Finke is essentially a bedrock gorge occupied by
a thread of intermittently-moving sediment which occurs as
deposits at the edge of the channel and bedforms, some of
which are hundreds of metres long. The frequency with which
these sediments move depends on the depth and velocity of

flow, the width of the gorge and the size of material which
varies from very fine sand through to cobbles and boulders.
Finer material, which travels as washload, is only found
in small quantities and is stored mainly within coarser
deposits. Very little weathered sediment occurs in the
active area of the gorge indicating that most features are
transient. The length of time they survive, however,
depends on the time between major floods which, in the last
fifty years or so, has been fairly short. It is therefore
likely that the morphology of more active areas of the gorge
is recent.

Several patterns of channel morphology or bedform complexes recur throughout the gorge. In the reach between
Palm Valley and the Ellery Creek junction, the bedrock gorge
is narrow and large flows are backed up by a series of tight
bends. The gradient is lower and much of the sand travels
as bedload producing what has been termed narrow gorge
morphology. The reaches between the northern end of the
gorge and Palm Valley and downstream from the Ellery Creek
junction are much wider and have a steeper gradient. Less
sand travels as bedload through these reaches and gravel
bedforms are more common. This produces a characteristic
wide gorge morphology.

A.   Narrow Gorge Morphology

An idealised representation of the bedforms and
sedimentary structures typical of narrow gorge morphology
is shown in Figure 11. The bed of the main channel is an
area of active sand transport and contains dunes, linguoid
bars or dunes over linguoid bars which are active during
most significant flows. While sand is the most common bed-
material, there are some areas of gravel. These occur in
long straight sections between bends where the channel is
less constricted and flow velocity is higher. Bedrock may
also be exposed in places. Slightly above the main channel,
extensive sand sheets occur which may have bed forms such
as linguoid bars but are frequently plane. These sheets
may occur as a central bar, dividing the channel, or at
either side. They are particularly well developed at or
on the downstream side of bends and may extend for several
hundred metres. Above the sand sheets and at the inside
of bends, point bars may occur. These features are not
well-developed, suggesting that once a significant stage
is reached, the channel is sufficiently constricted for flow
on the inside of bends to be erosional rather than
depositional.

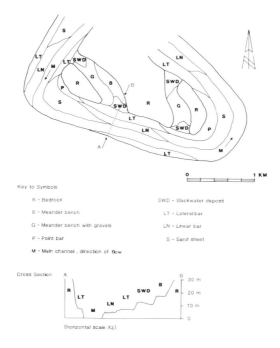

FIGURE 11. Idealised narrow gorge morphology.

Another kind of bar which occurs up to about the same height as point bars is the linear bar. These are sand with some gravel, occur in mid-channel, and are up to 5 m in height, about 10-20 m wide and may be more than 100 m long, although most are shorter. The linear bars are sometimes occupied by very large river red gums (*Eucalyptus camaldulensis*) suggesting that they pre-date the large floods of the 20th century, although most have layers of fresh sand on them. They have been interpreted as bed forms created as threads of sand move along the channel during the largest events and their survival may be closely related to colonisation by vegetation.

The Finke does not have a floodplain as such. There are, however, narrow flat-topped lateral bars which mainly occur in straight reaches at a variety of elevations above the active channel. These bars contain the finer sediments which might be expected in a floodplain but are now transient, many being actively destroyed by undercutting or having minor channels scoured between them and the bedrock

edges of the gorge.   No information is available on the age
of these bars but the sediments contained in them appear
fresh.   These are seen as an incipient floodplain created
during the long period between major floods (see below) and
now being removed by recent very large flows.

B.   *Wide Gorge Morphology*

Wide gorge morphology (Fig. 12) is best developed
downstream from Merrick Gully where the Finke crosses a
series of strike valleys and is not so deeply entrenched
as upstream.   There are, however, still occasional constrictions where the river passes through a strike ridge.
In these locations, the river usually consists of a wide,
heavily armoured single gravel, cobble, or even boulder-lined
channel with a large linear bar complex made up of sand and
gravel at one side. Away from the constrictions, the channel
may widen by a factor of 2-3.   In these wide reaches, it
takes on a braided form with two or more active gravel
channels with well-developed and regular bedforms including
gravel dunes at some locations.   The individual channels
are separated by large braid bars made up of sand and gravel
and in quieter locations, sand only.   These bars have been
active recently, given their fresh appearance and the age
of the vegetation occupying them.   The material at the edge
of the channel varies in age and stability.   In some areas,
fine material has accumulated to form a layered levee   or
incipient floodplain surface.   The upper part of this surface
shows recent activity in some areas in the form of scour
holes or mounds of fresh sand.   Elsewhere, it seems stable
and is probably a large slackwater deposit (see below)
formed between the edge of the main channel and the fringing
valley wall.   Beyond the slackwater deposits and levees,
the river is fringed by red sand dunes which overlay
weathered bedrock or ancient fluvial or colluvial material.
These dunes sometimes reach the edge of the channel and have
river frontages which have been trimmed during large floods.
     While most parts of the wide-gorge channel show evidence
of recent activity, there are apparently inactive cobble
or gravel benches which lie above the main river in a few
locations.   The benches are fluvial in origin, having bars,
channels etc. on them but are occupied by large corkwood
(*Hakea suberea*) trees of considerable age.   The gravel
benches could either be a terrace or more likely, the result
of a past flood which greatly exceeded the major events of
the past sixty years (see below).

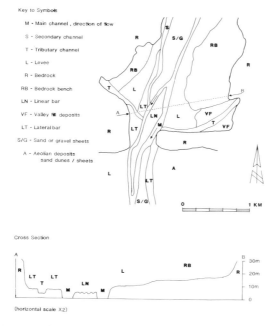

FIGURE 12.   Idealised wide gorge morphology

C. Slackwater Deposits

The Finke gorge is highly suited to the development and preservation of slackwater deposits. It experiences extremely large flows which carry very high sediment loads. It is in an arid area where the potential for erosion of the deposits in the period between floods is limited once they have become vegetated. The areas where the palaeomeanders intersect the main gorge are excellent environments for sedimentation, and finally most tributary catchment areas are small and do not produce enough flow to remove deposited material.

The sedimentology· of the Finke slackwater deposits is very simple. Most contain three or four layers of very uniform fine and very fine sand up to several metres in thickness. Little or no stratification is present within the individual units and there are no flow structures or tendency to fine upwards. This suggests that the deposits were laid down fairly quickly and there was little subsequent re-entrainment or reworking.

Three main types of slackwater deposit occur (Fig. 13). The first type consists of a mound occupying the mouth of a tributary valley with a steep gorge frontage and a gradual

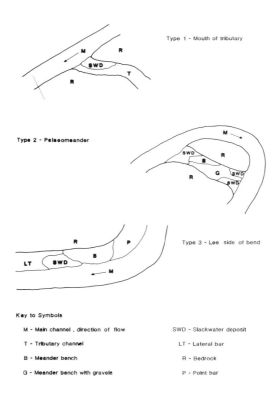

FIGURE 13.   *Types of slackwater deposit in the Finke Gorge.*

decline in elevation up the tributary. This type is always trenched and partly destroyed by the tributary stream and frequently tends to be lower and contain fewer of the oldest layers than other types. The mechanism of formation is probably an eddy in the tributary mouth where most of the deposition occurs and a limited amount of surging up the tributary. The top of the deposit is lower than other types because flow velocities in the eddy are sufficiently high to maintain some transport and to prevent deposited material from building up close to the water surface. The second type of deposit occupies both ends of each of the palaeo- meanders and may have a terrace or berm-like form with a flat top. There may also be a few low mounds on the flat top. Most of these deposits lie on top of bedrock and fringe the gravel deposits of the palaeomeanders. Others partly bury the gravels. Few have been extensively damaged by tributaries because the local catchment area of the palaeomeanders is small. The flat topped deposits are probably created by the surging of sediment-rich flows into a dead zone rather than eddying. The third type of slack-water deposit tends to occupy the bedrock shoulder of the main channel on the lee side of bends. It is probably almost entirely the result of one or more large eddies and varies in size with the configuration of the bend and the size of the sheltered zone. The largest of these deposits occur in the lee of the first bend downstream from the Ellery Ck. junction where a wide shoulder at the lowest level of the palaeomeander channel has been cut. This shoulder is partly covered by the fossil gravels found throughout the palaeomeander system but is sufficiently low to be almost entirely buried by a slackwater deposit hundreds of metres long.

V.   ESTIMATING THE MAGNITUDE AND FREQUENCY OF FLOODS FROM SLACKWATER DEPOSITS

A. Procedures

Given such an array of slackwater deposits, it is possible to estimate the magnitude of the largest floods which have occurred in a particular period and to reconstruct partial sequences of flood events. This can then be used to estimate the frequency of the largest events, a procedure which is fraught with difficulty using conventional flood estimation procedures based on the statistical analysis of flood records or rainfall-runoff modelling. The techniques of flood estimation from slackwater deposits are described elsewhere in the literature (eg. Baker, in press). Here concentration is on the special characteristics of the Finke slackwater sequence, how estimates of flood magnitude were obtained from it, and what the geomorphic implications of those floods are.

The Finke Gorge slackwater deposits are the best sequence so far discovered in Australia. They occur right through the reach upstream of the James Range anticline and occupy a wide range of locations. The very high degree of preservation occurs because they have been deposited at the entry and exit to each palaeomeander and on the bench at the edge of the present gorge which corresponds with the palaeomeander floor. These areas rarely have tributary streams of any significance crossing them so the slackwater deposits are not destroyed by tributaries breaking through and downcutting after flow in the main gorge has receded.

The existence of a well-preserved sequence of deposits over
a long reach of channel makes it possible to estimate the
gradient of the floods. It also allows verification of the
dates and stratigraphy obtained from individual deposits
which reduces the difficulties which sometimes occur when
trying to identify individual layers. This can sometimes
increase the amount of data available because some deposits
may not record all floods. Thus a variety of sedimentary
settings is an advantage.

The first step in estimating the magnitude of a sequence
of flood events from a slackwater deposit is to determine
its stratigraphy. In the slackwater deposits corresponding
with the highest flood levels on the Finke, individual
strata seem to represent individual floods and consist of
massive, relatively uniform layers up to 1 m in thickness
laid on top of each other (Fig. 14). These layers have to
be differentiated on the basis of subtle differences in
colour or former surfaces identified by burnt layers,
cobbles brought in by Aboriginals for tool making and, very
occasionally, hearths.

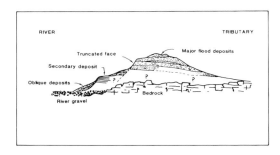

FIGURE 14. *Schematic diagram of slackwater deposit
stratigraphy.*

Once a flood puts down a layer of sediment on a
slackwater deposit, only higher floods can add material to
the top so the information is censored by a progressively
rising censoring level. Smaller floods can add material
to other parts of the deposit providing further information
on the flood sequence (Fig. 14). In the Finke, these
secondary deposits may occur as lower benches emplaced in
the trenches cut through the main slackwater deposits by
tributaries. They are, however, likely to be short-lived
in these locations. They can also form benches at the
front edge of the main deposit or be splayed up its lower
slopes as an oblique layer. Their life in these locations

may be short and they can be destroyed by the next large flood. The sedimentary deposits of the Finke have not been investigated in detail. Thus it is not known whether the information they contain is valuable for identifying the flood sequence. The one to one relationship between layers and flood events which occurs in the main slackwater deposits may not be the case in some of the secondary deposits. Small multiple layers are often present and there is sometimes evidence of slumping from the face of the main deposit. Caution should therefore be exercised when attempting to build a flood sequence from the stratigraphy of secondary deposits.

The elevation of the various strata within a slackwater deposit provide information on the *minimum* flood stage. The estimation of discharge from these stages requires data on channel cross-sections which may be obtained by survey, roughness coefficients which have to be estimated, and slope. A step-backwater analysis is then carried out using commercially-available computer programs such as HEC-2 (Hydrologic Engineering Centre, 1982) which provide information on velocity and slope as well as the stage-discharge relationship at each cross-section. Slackwater deposit heights may then be converted to discharge by determining the maximum flow represented by a particular stratum in the downstream sequence. Palaeoflood stages determined by step-backwater analysis are still subject to error. Roughness coefficients have to be estimated and, in the absence of other information, are usually assumed to remain constant along a reach and with varying discharge unless gauging data allows calibration (eg, Baker and Pickup, 1987). It is also necessary to assume that channel cross-sections remain largely as surveyed and have not scoured or filled significantly since the flood occurred. This is not always the case but the error can be limited by careful selection of cross-sections.

B. *Flood Discharge in the Finke*

Seven major floods occurring over the last 700 years were identified from a sequence of nine slackwater deposits upstream of Junction Waterhole (Table I). Six of these could be traced downstream of the Waterhole but, since only the three highest deposits were examined, it is not surprising that one was missed. Details of dates, site locations and stratigraphy are given by Webb *et al*, (in prep). Maximum discharges estimated from the step-backwater analysis using a roughness coefficient of .04 indicate that the 100

year flood has an estimated magnitude of close to
4000 m³s⁻¹ in the upper reaches of the gorge. This discharge
rises to 6000-7000 m³s⁻¹ below the junction with Ellery
Creek. The largest event recorded in the slackwater
sequence was greater than 5700 m³s⁻¹ which is large for a
4500 km² catchment by any standards. While such a flow falls
well below the envelope curve for maximum world floods, it
is of a similar magnitude to the largest events recorded
in more humid areas of Australia including the tropical
cyclone belt (eg, Pickup, 1976; Baker and Pickup, 1987).

The estimates of flood magnitude are of some importance
for flood estimation in central Australia where few gauging
records exist before 1950. These records span the wettest
period since European settlement so they are likely to
provide overestimates of both the magnitude and frequency
of flooding should future climatic conditions reflect those
of the past. Therefore the Finke data have been added to
the regional flood frequency curve for rock catchments
derived by MacQueen (1978) (Fig. 15). The plotting positions
assume that all the major floods have been identified, i.e.,
the record is not censored, and the discharges are correct
(see Baker and Pickup, 1987, for a test of accuracy of
discharge determination from slackwater deposits).

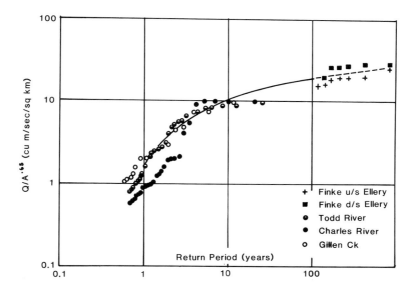

FIGURE 15. *Regional flood frequency curve for central
Australia (from MacQueen, 1978) incorporating floods
identified from slackwater deposits.*

The results suggest some fairly inspired (or lucky) extrapolation by MacQueen in deriving his curve from very short records. They also provide information in the crucial 100-1000 year range of return periods for which many hydraulic structures such as main road bridges are designed.

VI. SOME IMPLICATIONS

The record of major floods obtained from these slackwater deposits raises some questions about how arid zone rivers behave. Conventional models derived from small mid-latitude alluvial streams suggest that channels adjust their morphology to flow events of intermediate magnitude and frequency. Larger discharges dissipate their energy by flowing overbank and do not occur frequently enough to have a significant effect. Rivers confined in bedrock gorges cannot escape their channels so they flow at high velocities and stages which leads them to dominate morphology. Additionally, on rivers like the Finke, there may not be many events of intermediate magnitude as in humid rivers and they are not of long duration. Such events do occur in the headwater regions but do not travel far downstream because of high infiltration losses into the river's sandy bed during its long track across the Missionary Plain. For a flood to occur which passes along the full length of the river, heavy rainfall over a prolonged period and over a very large area is required. This means that when there is a flood, it is likely to be a large one and, because such events are rare, their effect on channel morphology is long-lasting.

The long term effect is not confined to the mountain or gorge sections of arid zone rivers. Studies of the flood plains of the Todd and Ross Rivers in the eastern MacDonnell Ranges (Patton and Pickup, in prep.) indicate the presence of palaeochannels more than five times the width of the modern river. Deposits in these suggest that they have only been fully active two or three times. Other parts of these floodplains contain huge ripple trains similar to but somewhat smaller than those described by Baker (1978) from the Lake Missoula floods. They are present on other floodplains and many alluvial fan systems throughout central Australia and provide strong evidence for overbank flows of very high magnitude occurring over large areas. These flows were, incidentally, significantly higher than the 1974 flood which is the largest detected for for the Upper Finke.

The very large discharges observed on the Finke add to the debate on the channel forms on Cooper Creek in the Channel Country of S.W. Queensland, one of the few other arid zone rivers to receive the attention of geomorphologists. The Cooper has two sets of multiple channel systems: a very large braiding pattern which can occupy the whole floodplain and a much smaller anastamosing pattern which is occupied by low and moderate flows. In the braided system, the bars are made up of mud and the pattern was formerly interpreted as a relict feature created by mud deposition on a braided sandsheet abandoned at least 20,000 years ago (Wasson, 1984) and possibly 50,000-100,000 years ago (Rust and Nanson, 1986). More recently, the floodplain has been seen as a feature of the current regime (Rust and Nanson, 1986; Nanson *et al*, 1986). They have identified a braiding mechanism in mud, have noted that water does flow through the braided system in large flows, but have not yet demonstrated movement on the braid bars and channels over time which should be expected if the braids are a product of the modern flow regime. The evidence for very large flows on the Finke supports the idea of active braiding on the Cooper because the braids could be created during extreme events and maintained for long periods in between (Nanson *et al*, this volume).

Another implication of the Finke data lies in their significance for climatic change. Of the seven major floods observed, four have occurred in the last seventy years and three in the last twenty. Such a clustering of events may be due to chance - a distinct possibility given the small sample. It could also mean that the slackwater sequence has not preserved the full historical record. If these suggestions are not correct, two other possibilities occur. The first one may be that the large flows occurred because runoff was high from large areas of bare soil in the catchment, the lack of vegetation being due to the effects of drought and grazing. The second explanation is more ominous. It could be that the changes in rainfall patterns reported in Australia and ascribed to the build-up of atmospheric CO_2 (eg, Pittock, 1981) not only involve shifts in the average. They could also mean increases in the magnitude and frequency of extreme events. If this is the case, flood frequency analysis based on historic rainfall and runoff data, in which the future is expected to mirror the past, could well become a flawed technique. It is therefore, perhaps, advantageous that most streamflow records in Australia are short and consequently biased towards the recent period with its extreme events (Erskine and Warner, this volume).

TABLE I. Discharges for Finke River Palaeofloods
Calculated from Step-Backwater Analysis and Slackwater
Deposit Stratigraphy.

Layer	Date	ABOVE ELLERY CK. JUNCTION Discharge (m^3s^{-1})	BELOW ELLERY CK. JUNCTION Discharge (m^3s^{-1})
1	1974	5700	8400
2	1972	3900	7900?
3	1967	3700	7900?
4	1921?	4700	8000
5	450 BP	4600	-
6	700 BP	4600	5900
7	850 BP	4400	8500

REFERENCES

Baker, V.R. (1978) Large-scale erosional and depositional
    features of the Channeled Scabland. In V.R. Baker and
    D. Nummedal (eds.) *The Channeled Scabland*, Planetary
    Geology Program, NASA, 81-115.
Baker, V.R. (in press) Palaeoflood hydrology and extra-
    ordinary flood events. *J. Hydrol.*
Baker, V.R. and Pickup, G. (1987) Flood geomorphology of
    the Katherine Gorge, Northern Territory, Australia.
    Geol. Soc. Am. Bull., 98, 635-646.
Baker, V.R., Pickup, G. and Polach, H.A. (1983) Desert
    palaeofloods in central Australia. *Nature*, 301, 502-504.
Callen, R.A. (1977) Late Cainozoic environments of part
    of northeastern South Australia. *J. Geol. Soc. Aust.*,
    24, 151-169.
Cook, P.G. (1968) 1:250,000 Geological Series - Explanatory
    Notes. Henbury, N.T. Bureau of Mineral Resources,
    Geology and Geophysics, Canberra.
Hydrologic Engineering Centre (1982) *HEC-2. Water Surface
    Profiles. Users Manual.* U.S. Army Corps of Engineers.
Mabbutt, J.A. (1966) Landforms of the western MacDonnell
    Ranges. In G.H. Dury (ed.) *Essays in Geomorphology*,
    Heinemann, N.Y., 83-119.

Machette, M.N. (1985) Calcic soils of the southwestern
    United States. In D.L. Wide (ed.) *Soils and Quaternary
    Geology of the Southwestern United States*, Geol. Soc.
    America Special Paper 203, 1-21.
MacQueen, A.D. (1978) Flood frequency in central Australia
    - a regional model. Water Division, Dept. of Transport
    and Works, Northern Territory.
Milnes, A.R. and Twidale, C.R. (1983) An overview of
    silicification in Cainozoic landscapes of arid central
    and southern Australia. *Aust. J. Soil Res.*, 21, 387-
    410.
Milton, D.J., Barlow, B.C., Brett, R., Brown, A.R.,
    Glikson, A.Y., Manwaring, E.A., Moss, F.J., Sedmik,
    E.C.E., Van Son, J. and Young, G.A. (1972) Gosses Bluff
    impact structure, Australia. *Science*, 175, 1199-1207.
Nanson, G.C., Rust, B.R. and Taylor, G. (1986) Coexistent
    mud braids and anastamosing channels in an arid zone
    river: Cooper Creek, central Australia. *Geology*, 14,
    175-178.
Pittock, A.B. (1981) Long term climatic trends in eastern
    Australia. In D.R. de Kantzow and B.G. Sutton (eds.)
    Cropping at the Margin, Aust. Inst. Agric. Sci. and
    Water Research Foundation, Sydney, 23-29.
Pickup, G. (1976) Geomorphic effects of changes in river
    runoff Cumberland Basin, N.S.W. *Aust. Geog.*, 13,
    188-193.
Quinlan, T. and Forman, D.J. (1968) 1:250,000 Geological
    Series - Explanatory Notes. Hermannsburg, N.T.
    Bureau of Mineral Resources, Geology and Geophysics,
    Canberra, 19pp.
Rust, B.R. and G.C. Nanson (1986) Contemporary and palaeo-
    channel patterns and the late Quaternary stratigraphy
    of Cooper Creek, southwest Queensland, Australia.
    Earth Surface Processes and Landforms, 11, 581-590.
Twidale, C.R. (1983) Australian laterites and silcretes:
    ages and significance. *Rev. Geol. Dynam. Geogr.
    Phys.*, 24, 35-45.
Twidale, C.R. and Milnes, A.R. (1983) Slope processes
    active late in arid scarp retreat. *Zeit. Geomorph.*,
    27, 343-361
Wasson, R.J. (1984) Field trip guide. *2nd Australian and
    New Zealand Geomorph. Conf.*, Broken Hill.

10

Episodic Changes of Channels and Floodplains on Coastal Rivers in New South Wales

Gerald C. Nanson

Department of Geography
University of Wollongong
Wollongong, NSW

Wayne D. Erskine

School of Geography
University of New South Wales
Kensington, NSW

I.   INTRODUCTION

Most modern commentary on alluvial rivers argues that they are in various states of equilibrium, where erosional forces are roughly balanced by those resisting the work of the river, and where channel and floodplain erosion is roughly equivalent to deposition (Richards, 1982; Knighton, 1984). However, a few studies in the United States have shown certain rivers can widen their channels and erode their floodplains catastrophically (eg Schumm and Lichty, 1963; Burkham, 1972) as in Schumm's (1973) geomorphic threshold and complex response model. Research has shown that NSW rivers can also adjust in a dramatic and complex way (Page, 1972; Henry, 1977) both to secular rainfall changes that have periodicities measured in decades (Erskine and Bell, 1982; Erskine, 1986a; Warner, 1987a; 1987b) and to catastrophic events that are spaced hundreds of years apart (Nanson and Hean, 1985; Nanson, 1986; Erskine, 1986b). These latter, much less frequent events seem only to affect portions of rivers where channels and floodplains evolve to a point where they become erosionally vulnerable (Nanson, 1986) or where the channels and floodplains are laterally confined by

erosion-resistant materials (Erskine, 1986b). The shorter
term climatic shifts are widespread and can affect much
longer lengths of river (Erskine, 1986a; 1986b; Warner, 1987a;
1987b). Such adjustments indicate conditions where, instead
of long-term stability, the landform fluctuates between two
(possibly more) different morphological states. The frequency
of abrupt response to threshold conditions is probably a
complex function of the temporal variability of the erosional
stresses involved (caused by variable precipitation and
associated runoff) and the temporal and spatial changes in
resisting stresses of the channel and floodplain boundary.
These resisting stresses are largely controlled by spatial
variations in sediment texture and boundary vegetation or by
progressive changes (both spatial and temporal) in channel
geometry and associated bankfull-discharge relationships
(Erskine, 1986a; Nanson, 1986, Fig. 10).

   This essay reviews recent evidence of episodic channel and
floodplain changes on NSW coastal rivers (Fig. 1 for
locations), particularly in the light of theories of landscape
evolution. The evidence presented here contradicts, in part,
the model of Wolman and Miller (1960) that states geomorphic
form is largely determined by processes of low magnitude and
high frequency.

II. EARLY INVESTIGATIONS OF NSW COASTAL RIVERS

   The erosional characteristics of these rivers have been
the subject of controversy since Darwin (1876), who visited
Australia in 1836, adopted the ideas of Charles Lyell and
ascribed the gorges of the Blue Mountains to marine action.
In contrast, Dana (1850), who visted the region in 1839,
clearly recognised the importance of contemporary flooding in
shaping these canyons. In the Kargaroo Valley he comments
on trying to conceive of the power available in a
flood 35 feet deep for a brook that is normally very
placid.

   Despite Dana's observations, the spectacular gorges of
southeastern Australia were long attributed to fluvial
activity during the Pleistocene. This conclusion was based
on two important misconceptions. Firstly, major uplift of
the Eastern Highlands was thought to have occurred in the
Pliocene or early Pleistocene and therefore there would not
have been much time for denudation (for example, Andrews,
1903). Secondly, contemporary river flows were seen to be of
little geomorphological significance.

   "It may be safely assumed that the excavation of the inner

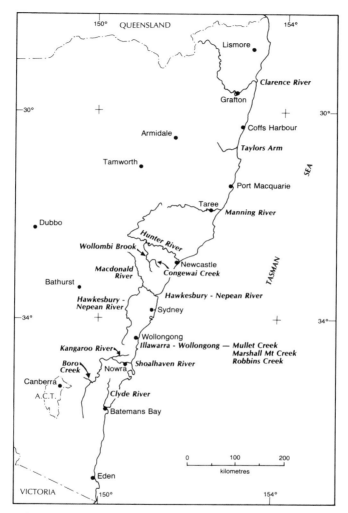

FIGURE 1. Location map.

gorges was accomplished almost entirely in the Pleistocene when, owing to heavier rainfall, the rate of erosion was very much greater than at present..." (Browne, 1969, p.566).

Amongst the earlier researchers, only Craft (1932) argued for the great antiquity of this landscape; this is now well established (Wellman and McDougall, 1974; Young, 1977). They demonstrated that much of the region had taken on essentially its present-day physiography by the mid-Miocene, and some of it much earlier. Young (1983) has

shown that fluvial processes have prevailed in southeastern
Australia for tens of millions of years on lithologies that
are not unduly resistant and under an almost continuously
humid climatic regime. However, the rates of stream erosion
have been "anomalously slow". There is no evidence to
indicate that the climatic fluctuations of the Pleistocene
produced undue erosion outside of the areas directly affected
by glacial or periglacial processes. Pleistocene river
terraces often have very widespread Tertiary counterparts;
an example on the Hawkesbury-Nepean River is the Londonderry
Terrace (Walker and Hawkins, 1957).

        Coastal NSW is an environment where most rivers have,
since the early to mid-Tertiary, operated uninterrupted by
the direct effects of glaciation or by prolonged periods of
aridity. Unlike many rivers studied elsewhere, these have
no large stores of sediment derived from Quaternary
glaciation or tectonism. Long-term denudation rates are
very low (Young, 1983) as are contemporary sediment transport
rates (Rieger and Olive, this volume). These rivers must
be close to being in a state of truly long term "adjustment"
with the landscapes they have created. However, this does
not necessarily mean they are under conditions of steady-
state or dynamic equilibrium. It remains to be established
just what the contemporary pattern of fluvial erosion and
deposition is in landscapes of such great antiquity,
relative stability and low denudation.

III.   CONCEPTS OF LANDSCAPE EVOLUTION

        Since the 1950s, most research on river systems has
been based on the concepts of steady state or dynamic
equilibrium (Fig. 2a and 2b) (Chorley *et al*, 1984). Such
systems are seen to exhibit a degree of homeostasis or
self-regulation whereby a change in one of the factors
governing equilibrium causes a slight change in the system
in a way that will minimise the original effects of the
change. The concept was not new; it was termed grade by
Gilbert (1877). Homeostasis has been so widely adopted as
the underlying paradigm in geomorphology that Chorley *et al*
(1984) were led to declare that it is characteristic of all
geomorphic systems in the long term, and most in the short.
        Channel cross sections, in particular, are seen to be
predictable and to result from a balanced set of conditions,
leading Richards (1982, p.148) to conclude that, "The
equilibrium thus results from an interaction between two
sets of variables - the force applied by the fluid and the

resistance to erosion mobilized by the sediment". Knighton (1984, p.97) argrees but is considerably more circumspect when he says, "Rivers that erode their boundaries flow in self-formed channels which, when subject to relatively uniform controlling conditions, are expected to show some consistency of form, at least on average".

The development of ideas on an equilibrium channel form led to the concept of there being a dominant channel-forming discharge which, more than any other flow, is responsible for maintaining a relatively uniform channel size and floodplain elevation (Wolman and Leopold, 1957; Wolman and Miller, 1960). According to this concept, while large floods may individually transport considerable sediment loads and alter significantly fluvial morphology, they occur too infrequently to have a long term effect on the channel. Bankfull flow was seen to have particular morphological significance for it seemed logical that channels adjusted to flows that just filled the channel cross section. Excess flow routed across the floodplain should have little bearing on channel morphology. Channels in a steady state equilibrium (Fig. 2a) exhibit a statistical equilibrium whereby the channel state does not change over time. This does not mean that channel geometry is invariant but that it oscillates about an average condition which is adjusted to bankfull discharge.

Numerous authors have presented evidence demonstrating the morphological significance of bankfull discharge including Pickup and Warner (1976) for streams in the Cumberland Basin near Sydney, and there is little doubt that the theory applies to many rivers world wide. However there is a growing body of evidence that shows that, in certain environments, extreme erosional events are geomorphically more significant than are numerous moderate events (Hack and Goodlett, 1960; Anderson and Calver, 1977). Even in their original paper, Wolman and Miller (1960) recognised that streams in small basins, particularly in arid environments, could trasport large amounts of total load during infrequent high-magnitude events, a point later amplified by Wolman and Gerson (1978). Pickup and Rieger (1979) concluded that channel form is the weighted sum of effects of all competent events.

Paradoxically, it was Andrews (1907), so ready to invoke exceptionally different past climates to explain the present landscape (Andrews, 1913), who was amongst the first in NSW to propose that "Titanic forces" associated with very large contemporary floods are instrumental in causing sustained landscape change. In fact, independently both he and Gilbert argued that quite possibly "the effect of the

maximum flood of the decade or generation or century
surpasses the combined effects of all minor floods.   It
follows that the dimensions of the channel are adjusted to
the great flood and adjusted to its needs" (Gilbert, 1883-4,
p.90).

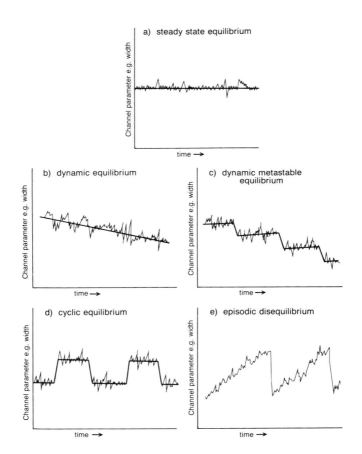

FIGURE 2.   Models of equilibrium (explanation in text).

    In 1973 Schumm proposed that geomorphic systems could
episodically change dramatically from one morphological
condition to another, when the system exceeds a critical
threshold in the balance between the landform's ability to
resist morphological change and forces acting to initiate
change.   He showed that equilibrium conditions can prevail

but are distinctly different on each side of the threshold
(Schumm, 1977). This situation is described by Chorley and
Kennedy, 1971; p.202) as dynamic metastable equilibrium for
it involves progressive long term changes in morphology or
energy in a particular direction, but with much of the change
taking place at the threshold points (Fig. 2c). Yet in some
natural systems there are two additional types of change
which here are termed cyclic equilibrium, where a new
equilibrium condition is approached between each threshold
of change (Fig. 2d) and episodic disequilibrium (Fig. 2e).
In the latter there is no evidence of distinct equilibrium
conditions at any point, although the rate of change between
each threshold is much slower than at the threshold. These
two oscillating states occur within the graded time scale
(10^0-10^3 years) of Schumm (1977) and provide the focus here
for explaining channel and floodplain changes on the
coastal rivers of NSW.

IV.   EPISODIC GEOMORPHIC CHANGES OF NSW COASTAL RIVERS

    While a number of engineering studies have documented
considerable stream and floodplain erosion in coastal NSW
in the last 3-4 decades (eg Reddoch, 1957), Page (1972) was
probably the first to recognise the geomorphic significance
of erosional thresholds. His study of Wollombi Brook (Fig.
1) described considerable channel erosion and floodplain
deposition along part of this stream resulting from the June
1949 flood. By the early 1970s Page observed what he
thought was the beginning of channel recovery, an observation
that would imply that such floods are part of a cyclical
sequence, relatively infrequent in occurrence. Hickin and
Page (1971) recognised that the alluvium in the narrow
sandstone valleys of the Sydney Basin is vulnerable to
erosion by catastrophic floods.
    In a Macdonald River study, Henry (1977) presented
detailed descriptions of catastrophic channel changes
resulting from a series of floods between 1949 and 1955. He
was unaware of any secular regime shifts to explain such
changes. As a result he ascribed the channel changes to
the effects of numerous large to moderate floods in rapid
succession, citing Schumm and Lichty's (1963) and Burkham's
(1972) studies of semi-arid rivers as overseas examples of
similar flood-induced channel changes. However, the rivers
studied by Page and Henry are located in a region of
distinctly humid climate. These initial studies provided a

different perspective for the study of NSW rivers. The
dynamic equilibrium paradigm could no longer provide a
complete explanation for all of the channel changes observed.
On the basis of work done over the last 25 years, it now
appears that these rivers respond both to extrinsically and
intrinsically controlled thresholds (Schumm, 1977).

V. CHANNEL CHANGE DETERMINED BY CLIMATIC (EXTRINSIC) FACTORS

A. *Cyclical Equilibrium Changes*

Following Kraus' (1955) initial observations, Pittock
(1975) and Cornish (1977) have shown that there has been
a statistically significant increase in mean annual rainfall
in southeastern Australia since the mid 1940s. Cornish
found the increase was greatest in summer. Pickup (1976)
demonstrated that this increased rainfall resulted in an
increase in the magnitude and frequency of floods in the
Cumberland Basin, immediately west of Sydney. In an
important development, Pickup linked this wet period to a
commensurate increase in channel erosion and bedload
transport. Extending this theme of relating secular climatic
change and geomorphic response, Erskine and Bell (1982)
undertook a detailed study of the upper Hunter Valley (Fig.
1). They showed convincingly that the post-1946 wet period
coincided with increases in width, hydraulic capacity and
bedload transport of these channels. Hean and Nanson ·
(1987) calculated that the increase in discharges on the
Shoalhaven River since the mid-1940s has increased bedload
yields by up to 60-100%. Although human disturbance in
coastal NSW has been cited as the main cause of increased
flooding and/or accelerated erosion (eg Maiden, 1902;
Monteith, 1953), Pickup (1976) and Erskine and Bell (1982)
show that meteorological variations can greatly overshadow
the effects of changed landuse practices.

In another study of the Macdonald, Erskine (1986a)
extended the work of Henry (1977), highlighting the channel
metamorphosis that resulted from the period of increased
rainfall from 1947. Erskine showed that channel changes
resulted, not from the chance occurrence of several major
floods as suggested by Henry, but from a period of
persistent climate change. Substantial increases in width
and bed-material transport were found as well as decreases
in depth, sinuosity and the proportion of silt and clay in
the channel boundary. With the advantage of longer
observations, Erskine was able to re-examine Henry's

suggestion of later channel recovery. It was shown that the Macdonald remained unstable from 1955 to at least the early 1980s, particularly in relation to erratic bed elevations.

In a study of the Nepean River near Penrith, Warner (1987a; 1987b) described the periods of higher than average and lower than average rainfall and flooding, as flood-dominated regimes (FDR) and drought-dominated regimes (DDR), respectively. These tend to last from 3 to 5 decades and FDR have mean annual flood discharges from 2 to 4 times greater than DDR (Warner, 1987a; 1987b; Erskine, 1986a). Within the historical record, DDR prevailed from about 1821 to 1856 and again from 1901 to 1948, whereas FDR occurred from 1799 to 1820, 1857 to 1900 and 1949 to the present. Using a set of channel survey data started in 1862, Warner (1987b) has shown that during FDR channel widths generally increase and depths decrease, whereas in DDR, these changes are reversed. As a result, he argues that in certain erodible reaches there may well be two morphologically distinct channels; an inner, smaller one related to the DDR and a larger one, flanked by high eroded alluvial banks, relating to the FDR. As noted by Pickup (1986), channel response to alternating FDR and DDR is greatest on sand-bed and bank reaches (his sand zone). This is hardly surprising as the sensitivity to change of rivers is a function of the distribution of resisting and erosive forces (Brunsden and Thornes, 1979) and the resisting forces are lowest in sand-rich channel perimeters. However, as Warner's (1987a; 1987b) work demonstrates, channels with more resistant boundaries in Pickup's (1986) gravel-sand transition, armoured and even bedrock zones, have responded albeit slowly to shifts in flood regime. Sink zone channels appear the most insensitive to regime shifts because of cohesive boundaries, (Warner and Paterson, 1987) often well vegetated, with low in-channel stream power induced by gentle slopes and small channel capacities (Nanson and Young, 1981a; Nanson and Hean, 1985). It should also be stressed that these alternating flood regimes often result in temporal variations in the longitudinal distribution of bed-material zones. The Macdonald River sand zone has extended downstream by about 10 km into the sink zone since 1949, while the transition zone in Taylors Arm (Fig. 1) has advanced a similar distance over the same period. In DDR it appears that this remobilised sediment is exhausted and the channel slowly reverts back to the characteristics of its earlier bed-material zone.

This review of cyclical equilibrium channel changes on NSW coastal rivers shows that certain rivers can switch from narrow deep channels to wide shallow channels in response to extrinsic factors, namely DDR and FDR (Erskine and Warner, this volume). In both regimes, channels attempt to reach a type of fluctuating steady-state equilibrium, only to switch alternative equilibrium states in response to a change in climate. These episodes of distinctly different channel regime are fairly short-lived (30-50 years) and so channels probably never properly adjust to a clearly-defined equilibrium with the prevailing hydrological conditions (see Erskine and Melville's (1983) evidence of considerable short-term fluctuation in bed elevation on the Macdonald River). Oscillations between two quasi-equilibrium states are defined as cyclic equilibrium (Fig. 2d). It is different to the dynamic metastable equilibrium of Chorley and Kennedy (1971) because firstly, there is no detectable change in the average position of the equilibrium (such as that caused by long term denudation), and secondly, the thresholds alternate in their direction of change rather than progressing in the same direction (compare Figs. 2e and 2d).

B. *Infrequent Catastrophic Change*

   All rivers are subject to very large, infrequent events and some will respond with catastrophic changes in channel and floodplain morphology. Here catastrophic change is defined as a morphological change of great magnitude that occurs in a geomorphological system, on average, less frequently than about once every 200 years. This means that there would have been less than 50 such events during the entire Holocene and possibly one since European settlement. The geomorphological effects of such extreme events can exhibit great persistence but unless the features formed are observed in the making, it may not be obvious that they resulted from a single event.
   In order to investigate the role of catastrophic floods in channel formation two case studies are presented. The first examines the impact of a single catastrophic flood (an extrinsic event) in the Illawarra region south of Sydney (Nanson and Hean, 1985) while the second describes the response and recovery of Congewai Creek and Wollombi Brook in the Hunter Valley to a series of large floods between 1949 and 1955 (Erskine, 1986b; pp.153-194).

   1. Illawarra Streams. Small streams flowing off the Illawarra coastal escarpment near Wollongong, NSW (Fig. 1) exhibit a wide range of energy gradients in this humid environment (Nanson and Young, 1981a; 1981b). These are

ideal for the study of geomorphological responses to an
extreme flood event. Basin areas range from 20 to 40 km^2,
headwaters are very steep and densely forested and valley
sides confine channels which have almost no floodplains.
The foothills and coastal plain have broad low-gradient floodplains under dense pasture. Within the confined and steeply
sloping valleys of the escarpment and its foothills, channels
increase in size downstream in response to increasing basin
area. However, in the open valleys of the plain, they
dramatically reduce their cross-section area and bankfull
capacity, probably in response to an abrupt decrease in
channel gradient and associated stream power. Upstream
floodwaters have insufficient erosive power to maintain
channel capacities capable of containing even a small flood
(Nanson and Young, 1981a).

The upstream floodplains are narrow, relatively shallow
and composed of coarse basal gravels and boulders deposited
under high energy conditions, with a relatively thin
overburden of fines. In contrast, the downstream floodplains
consist of a very thick deposit of fine-textured overbank
sediment deposited under conditions of low energy. These
have been produced by the Holocene marine transgression
forming a large sink zone (Pickup, 1986) in the lower
sections of these valleys. A February 1984 storm with rainfall recurrence interval of 200-300 y or more provided near
maximum erosional stress conditions (Nanson and Hean, 1985).
About 800 mm fell, most of it in a 9 hour period, making
it the largest 24 hour fall in temperate Australia. Three
drainage basins were studied each with channel slopes of
0.01-0.025 in the foothills and 0.005-0.0005 on the coastal
plain. In Mullet and Marshall Mount Creeks, a large
proportion of the flood discharge was retained in the
upstream steep-gradient channels causing considerable
erosion; one cross section on Mullet Creek increased in area
by 229% and another by 338%. In contrast, the downstream
low-gradient, small-capacity channels displaced much of
their flow over the floodplains, and there was little channel
change (average of +20%). Similarly, the upstream floodplains experienced considerable scouring, the channel of
Forest Creek undergoing a total avulsion (Nanson and Hean,
1985). However, when changes in the cross-sectional area
of Mullet and Marshall Mount Creeks are regressed against
the calculated stream power there is a good, correlation
(significant at the 95% level), explaining 73% of the
variance (Nanson and Hean, 1985). Surveying of the upstream
channels in 1987 has shown that, rather than recovering to
their pre-flood condition, they have slightly enlarged. It

is reasonable to suppose that they will eventually return
to their pre-flood size, but contraction will be a very
slow process, perhaps dependent on moderate floods supplying
coarse debris from upstream, to start the process of
accretion.

This storm, acting as it did on both high and low energy
environments, has shown that the geomorphic response to an
extreme event is highly dependent on the potential energy
in the system and the nature of the channel boundary. The
upstream areas have been greatly altered and are not yet
showing any sign of significant recovery whereas evidence
of the storm downstream has been all but lost.

2. *Wollombi Brook*. The storm of 17-18 June 1949 was
the largest in Wollombi Brook in the Hunter Valley (Fig.
1) since at least 1827. The highest rainfall recorded was
508 mm but the lack of gauges in the 2000 km^2 catchment
precluded rainfall frequency analysis. However, peak
discharge near the basin outlet was almost 27 times the mean
annual flood but its return period was only 87 years on the
annual series. This is to be expected as Wollombi Brook
has one of the largest Flash Flood Magnitude Indices (Baker,
1977) ever recorded in Australia (Erskine, 1986b). Therefore
the slope of the annual series flood frequency curve is
exceptionally steep and the return period of large events
is relatively small in comparison to rivers with gentle flood
frequency curves (Baker, 1977). The 1949 flood was a
large event because its peak discharge for a basin of this
size has only been exceeded twice on all NSW rivers. A
series of large floods in rapid succession followed the 1949
flood. Between 1949 and 1955 only one gauge operated
continuously in the basin and 18 floods greater than the
mean annual were recorded in this period. It seemed
probable that the geomorphic effects of the 1949 flood
would be magnified by the repeated occurrence of large floods
immediately afterwards.

Many historical sources were used to document channel
response and recovery, if any, for these floods (Erskine,
1986b). Of the 130 km of sand-bed river investigated only
the lower 89 km were significantly impacted by these floods.
Rather surprisingly, the channel at the storm centre and
downstream did not change significantly. Further downstream,
the channel widened by about 100% and the eroded sand was
deposited in the channel bed, producing up to 4 m of
aggradation. The eroded downstream sections are laterally
confined by bedrock and terraces whereas the stable,
upstream sections are essentially unconfined by any erosion-
resistant materials other than finer grained floodplain

sediments. Channel recovery in the flood-impacted reach
is now well advanced. In-channel bench construction has
decreased the width and since about 1960 there has been up
to 1 m of degradation. Benches are recovery landforms and
temporary sediment storages constructed by small to moderate
floods within widened channels cut by the large event(s).
Hydraulic geometry changes are such that velocity for a given
discharge is increasing over time thus increasing the
river's capacity to evacuate the flood-aggraded sand.
Therefore, the rate of channel recovery should accelerate
until the next large flood (or series) initiates a new cycle
of channel enlargement. Channel recovery has been promoted
by a lack of large floods and the completion of extensive
river training works.

VI. CHANNEL CHANGE DETERMINED BY INTRINSIC FACTORS

This section describes the response of gradually
evolving channels and floodplains to intrinsic erosional
thresholds in the Clyde and Manning valleys (Nanson, 1986),
as well as the relevance of valley-floor slope thresholds
to gully erosion on Boro Creek, a tributary of the Shoalhaven
River (Fig. 1).

A. *Clyde and Manning Rivers*

It has been widely assumed that floodplains form largely
by lateral accretion of point-bar sediments, a process
keeping morphology in equilibrium with the flow regime
(Wolman and Leopold, 1957). Vertical-accretion sediments
are thought to prevail only along low energy and slow
sedimentation reaches. However, certain non-migrating
reaches of the Clyde and Manning Rivers (Fig. 1), form
floodplains by vertical accretion and erode them
catastrophically (Nanson, 1986). At the study sites the
rivers occupy rugged 400 and 6500 km^2 basins (respectively)
in areas of humid climate (annual rainfall up to 1300 mm).
Floodplains show considerable relief, are relatively narrow
and have well defined levees, sometimes 7 m above adjacent
backchannels. Floodplain scour channels indicate periodic
high-energy flooding. The stratigraphic architecture indicates
deposition by vertical accretion with no laterally sloping
point-bar strata.
   During a series of eight floods ranging from recurrence
intervals of 3.8 to 52 years (annual series) during the period

1968 to 1978, approximately half the 250 m width of the
Manning floodplain at Charity Creek was stripped down to
the basal gravels over more than 600 m. These floods were
not large and most of the erosion occurred before the 52
year flood, but in combination they caused massive local
scour which should have been more widespread than it was.
The uneven erosion pattern suggests that some parts of the
floodplains were more vulnerable to these floods than others.
   ^{14}C dates from basal sands of eroded floodplain at
Charity Creek show that the floodplain had probably been
eroded previously about 350 to 400a BP, and has since been
reformed (Nanson, 1986). At Bretti on the upper Manning
the floodplain has been cut and filled so as to produce
three distinct levels which look like a terrace, a floodplain and a bench. However, basal dates show the "terrace"
to be 490 ± 110a BP (ANU 5466) and the bench to be
indistinguishable from modern (ANU 5465). The Clyde River
floodplains yielded basal ages of about 2000a BP (Nanson,
1986).
   From this evidence, Nanson (1986) proposed a theory of
episodic stripping and vertical accretion for these floodplains in confined mountain valleys. Because the rivers
often abut bedrock at their outer banks they often fail to
laterally migrate, and as a result their floodplains build
vertically.

   "The continued growth of these floodplains brings about
their ultimate destruction. As the levee banks gain in
height, they form steep-gradient flood plains that confine
larger and larger flows to the main channel and to steep-
gradient backchannels on the floodplains. Eventually,
erosional thresholds, both in channel and on the floodplain,
are exceeded, and wholesale erosion results. What follows
is a gradual period of floodplain reconstruction by vertical
accretion" (Nanson, 1986, p.1474).

   The result is largely disequilibrium; the channel and
floodplain would almost never be in dynamic equilibrium
with the prevailing flow regime. It is either gradually
building and altering the flow conditions, or it is severely
eroding.

B. Boro Creek

   Boro Creek, a tributary of the Shoalhaven, is located
in the Southern Tablelands (Fig. 1). Three large

discontinuous, valley-botton gullies are present on the
upper reaches below the Mulwaree Fault (Erskine and Melville,
1984).   The basin areas at the upper and downstream limits
of the gullied section are 3 and 27 km^2, respectively.
Valley-floor slopes were measured to determine whether a
threshold valley-floor slope could be identified above which
the probability of incision is very high. Figure 3 shows
an almost perfect separation of gullied and ungullied reaches
on a bivariate plot of valley-floor slope versus catchment
area.   Only two points, one gullied and one ungullied plot
anomalously.   The latter has been trenched since the survey.
The gullied site reflects a former low angle fan trenched
by a knickpoint initiated on the steeper reach downstream.
The dashed line (Fig. 3) discriminates the critical
valley-floor slope above which the probability of
entrenchment is high.   A discriminant function was used to
assess the statistical validity of the two data groups.
This showed that the multivariate means of the two data sets
are significantly different ($p < 0.05$).   Steeper sections
often coincide with valley-floor constrictions where the
channel cuts through a minor resistant strike ridge.   As
noted by Schumm (1973), relatively slow and continuous
sedimentation increases valley-floor slope until an
instability threshold is reached.   Then the landform changes
from an unincised, often swampy, reed-covered valley floor to
a discontinuous valley-bottom gully.   Chains of ponds (Eyles,
1977) are characteristic of the ungullied valley floors.

VII. DISCUSSION AND CONCLUSIONS

The last 15 years have seen major advances in Australian
fluvial geomorphology.   This progress has come after strong
reliance on preconceptions of how Australian rivers should
behave based on North American and European literature
(Pickup, 1986).   These advances have been based on the
recognition of three fundamental aspects of the Australian
environment.   Firstly, the landscape is very much older than
previously suspected.   Secondly, denudation rates are very
low compared to the tectonically active or recently
glaciated areas.   Finally, Quaternary climatic change in
Australia has followed a very different path to that in the
temperate and cool temperate areas of the Northern
Hemisphere where most palaeoclimatic research has been
done.
Early work was based on the assumption that
geomorphological time was relatively short and that a high-

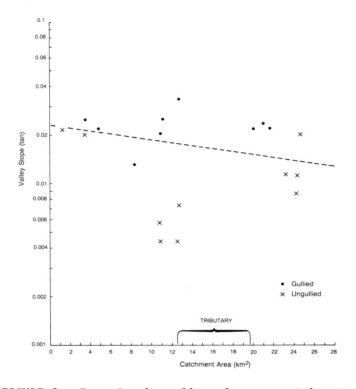

FIGURE 3.   Boro Creek: valley slope v. catchment area.

rainfall Pleistocene climate was primarily repsonsible for
much riverine morphology.   Paradoxically, it was Professor
Dury, who came to the University of Sydney from Britain via
the U.S.G.S., who was to precipitate a fundamental change
in the study of east coast rivers.   As Pickup (1986) notes,
fluvial geomorphology was largely unknown here until the
1960s, and Dury himself was to do little work on
contemporary rivers.   However, two of his students (Hickin
and Riley) and two other students (Page and Pickup) soon
identified the sometimes catastrophic and flood-dominated
character of the coastal rivers.   Another of Dury's students
(Young) recognised the geomorphological significance of
the great age of the adjacent landscapes.

Instead of abundant sediment transport, as might occur
in vigorously eroding landscapes, the rivers of coastal NSW
appear to be supply limited.   Roy and Crawford (1977) have
shown that few supply sand or gravel to the coast and Hean and
Nanson (1987) also argue for their limited supply. An
abundant sediment load would keep channels in a condition

of dynamic change as bars build and erode, banks collapse and point bars accrete, largely in response to low magnitude, high frequency events. There seem to be few, if any, rivers of this character in coastal NSW. Instead, they store sediment and release it in large, relatively infrequent pulses. The more dynamic sand-bed rivers, such as the Macdonald, store sediment in channel benches during DDR and erode these during FDR, thereby rapidly adjusting their channel geometry. This has been termed cyclic equilibrium as the channel oscillates between two different states. The less responsive rivers have gravel beds (like the Clyde and Manning), build their floodplains more gradually, then erode them catastrophically when the valley can no longer accommodate the obstruction of the growing floodplain. These catastrophic events may well be triggered during FDR, but as Erskine and Bell (1982) note, very large and infrequent storms are equally likely to occur during DDR. Because this system never seems to achieve a steady-state condition and is probably spatially and temporally less regular than the former system, it is referred to as episodic disequilibrium (Nanson, 1986).

Gully erosion is often initiated when a threshold valley-floor slope is exceeded (Schumm, 1973). Recent work in the Hunter (Melville and Erskine, 1986) outlined the control of slope on knickpoint initiation. Furthermore, Erskine (1986b), from detailed historical records and field surveys, showed that many discontinuous valley-bottom gullies in the upper Wollombi Brook basin start on steep and narrow valley-floor segments. Gully erosion caused by the exceedance of a critical slope is another example of episodic disequilibrium.

The Illawarra streams show that potential energy has a profound effect on the system's response. High energy streams will adjust catastrophically whereas those on lowland plains may never show such change. This marked difference in stream power is reflected in boundary materials, channel geometry and floodplain stratigraphy (Nanson and Young, 1981a; 1981b). The same energy/erosion relationships were found on Wollombi Brook (1949 to 1955), except there the low energy reaches are in the headwaters. Greater erosion occurred downstream because of the closer marginal confinement of the channel by bedrock valley sides or resistant terraces.

The pattern of fluvial erosion and deposition on the coastal rivers of NSW, a region of great antiquity, tectonic stability and slow denudation, is not one of simple equilibrium conditions. Instead, these rivers store their limited sediment load over prolonged periods and then release it almost instantaneously during infrequent large events. Many NSW coastal rivers besides those discussed above have channel forms largely determined by floods of high magnitude and low frequency.

REFERENCES

Anderson, M.G. and Calver, A. (1977) On the persistence of landscape features formed by a large flood. *Trans. Inst. Brit. Geog.,*   NS 2, 243-254.

Andrews, E.C. (1903) Notes on the geology of the Blue Mountains and Sydney Districts. *Proc. Linn. Soc. NSW,* 28, 786-825.

Andrews, E.C. (1907) The geographical significance of floods with especial reference to glacial action. *Proc. Linn. Soc. NSW,* 32, 795-834.

Andrews, E.C. (1913) Report on the Cobar Copper and Goldfields. Part 1. *Min. Res. N.S.W.,* 17.

Baker, V.R. (1977) Stream-channel response to floods with examples from central Texas. *Bull. Geol. Soc. Amer.,* 88, 1057-1071.

Browne, W.R. (1969) Geomorphology. *J. Geol. Soc. Aust.,* 16, 559-580.

Brunsden, D. and Thornes, J.B. (1979) Landscape sensitivity and change. *Trans. Inst. Brit. Geog.,* N.S. 4, 463-484.

Burkham, D.E. (1972) Channel changes of the Gila River in Stafford Valley, Arizona, 1846-1970. *U.S.G.S. Prof. Paper 655-G.*

Chorley, R.J. and Kennedy, B.A. (1971) *Physical Geography: A Systems Approach,* Prentice Hall, London.

Chorley, R.J., Schumm, S.A. and Sugden, D.E. (1984) *Geomorphology,* Methuen, London.

Cornish, P.M. (1977) Changes in seasonal and annual rainfall in New South Wales. *Search,* 8, 38-40.

Craft, F.A. (1932) Physiography of the Shoalhaven River Valley; Part IV, conclusion. *Proc. Linn. Soc. N.S.W.,* 57, 245-260.

Darwin, C.R. (1876) *Geological Observations on the Volcanic Islands and part of South America Visited During Voyage of H.M.S. Beagle,* Smith, Elder, London (2nd edn.).

Dana, J.D. (1850) On the degradation of rocks of New South Wales and formation of valleys. *Amer. J. Sci.,* 9, 289-294.

Erskine, W.D. (1986a) River metamorphosis and environmental change in the Macdonald Valley, New South Wales since 1949. *Aust. Geog. Studies,* 241, 88-107.

Erskine, W.D. (1986b) River metamorphosis and environmental change in the Hunter Valley, New South Wales. Unpubl. PhD thesis, Univ. NSW.

Erskine, W.D. and Bell, F.C. (1982) Rainfall, floods and river channel changes in the upper Hunter. *Aust. Geog. Studies.*, 20, 183-196.

Erskine, W.D. and Melville, M.D. (1983) Impact of the 1978 floods on the channel and floodplain of the lower Macdonald River, NSW. *Aust. Geog.*, 15, 284-292.

Erskine, W.D. and Melville, M.D. (1984) Sediment movement in a discontinuous gully system at Boro Creek, Southern Tablelands, NSW. In R.J. Loughran (compiler), *Drainage Basin Erosion and Sedimentation*, University of Newcastle and Soil Cons. Service of NSW, Newcastle, 197-204.

Eyles, R.J. (1977) Changes in drainage networks since 1820, Southern Tablelands, NSW. *Aust. Geog.*, 13, 377-386.

Gilbert, G.K. (1877) *Report on the Geology of the Henry Mountains*, US Dept. Interior, Wash. DC, 93-144.

Gilbert, G.K. (1883-84) The topographic features of Lake Shores. *U.S.G.S. 5th Ann. Rept.*, 77-123.

Hack, T.J. and Goodlett, J.C. (1960) Geomorphology and forest ecology of a mountain region in the central Appalachians. *U.S.G.S. Prof. Paper 347.*

Hean, D.S. and Nanson, G.C. (1987) Serious problems in using equations to estimate bedload yields for coastal rivers in NSW. *Aust. Geog.*, 18, 114-124.

Henry, H.M. (1977) Catastrophic channel changes in the Macdonald Valley, New South Wales, 1949-1955. *J. Roy. Soc. NSW*, 110, 1-16.

Hickin, E.J. and Page, K.J. (1971) The age of valley fills in the Sydney Basin. *Search*, 2, 383-384.

Knighton, D. (1984) *Fluvial Forms and Processes*, Arnold, London.

Kraus, E.B. (1955) Secular changes of east-coast rainfall regimes. *Quart. J. Roy. Met. Soc.*, 91, 430-439.

Maiden, J.H. (1902) The mitigation of floods in the Hunter River. *J. Roy. Soc. NSW*, 36, 107-131.

Melville, M.D. and Erskine, W.D. (1986) Sediment remobilization and storage by discontinuous gullying in humid southeastern Australia. *I.A.H.S.*, Publ. No. 159, 277-286.

Monteith, N.H. (1953) The contribution of historical factors to the present erosion condition of the Upper Hunter River. *J. Soil Cons. NSW*, 9, 85-91.

Nanson, G.C. (1986) Episodes of vertical accretion and catastrophic stripping: a model of disequilibrium flood plain development. *Bull. Geol. Soc. Amer.*, 97, 1467-1475.

Nanson, G.C. and Hean, D.S. (1985) The West Dapto flood of February 1984: rainfall characteristics of channel changes. *Aust. Geog.*, 16, 249-257.

Nanson, G.C. and Young, R.W. (1981a) Downstream reduction of rural channel size with contrasting urban effects in small coastal streams of Southeastern Australia. *J. Hydrol.*, 52, 239-255.

Nanson, G.C. and Young, R.W. (1981b) Overbank deposition and floodplain formation on small coastal streams of New South Wales. *Zeit. f. Geomorph.*, 25, 332-347.

Page, K.J. (1972) A Field Study of the Bankfull Discharge Concept in the Wollombi Brook Drainage Basin, New South Wales, Unpubl. MA (Hons.) thesis, Univ. Sydney.

Pickup, G. (1976) Geomorphic effects of changes in river runoff, Cumberland Basin, NSW. *Aust. Geog.*, 13, 188-193.

Pickup, G. (1986) Fluvial landforms. In D.N. Jeans (ed.) *Australia: A Geography Vol. 1, The Natural Environment*, Sydney Uni. Press, 148-179.

Pickup, G. and Rieger, W.A. (1979) A conceptual model of river channel changes with time. *Earth Surface Processes*, 4, 37-42.

Pickup, G. and Warner, R.F. (1976) Effects of hydrologic regime on magnitude and frequency of dominant discharge. *J. Hydrol.*, 29, 51-75.

Pittock, A.B. (1975) Climatic change and the patterns of variation in Australian rainfall. *Search*, 6, 498-504.

Reddoch, A.F. (1957) River control work in the non-tidal section of the Hunter River and its tributaries. *J. Inst. Eng. Aust.*, 29, 241-247.

Richards, K. (1982) *Rivers, Form and Process in Alluvial Channels*, Methuen, London.

Roy, P.S. and Crawford, E.A. (1977) Significance of sediment distribution in major coastal rivers, northern NSW. *Proc. 3rd Aust. Conf. Coastal and Ocean Eng.*, 177-184.

Schumm, S.A. (1973) Geomorphic thresholds and complex response of drainage systems. In M. Morisawa (ed.) *Fluvial Geomorphology*, Binghamton, Pubs. in Geomorph. N.Y., 299-310.

Schumm, S.A. (1977) *The Fluvial System*, Wiley, New York.

Schumm, S.A. and Lichty, R.W. (1963) Channel widening and floodplain construction along the Cimarron River in Southwest Kansas. *U.S.G.S. Prof. Paper 352-D*.

Walker, P.H. and Hawkins, C.A. (1957) A study of river terraces and soil development on the Nepean River, NSW. *J. Roy. Soc. NSW*, 91, 67-84.

Warner, R.F. (1987a) Spatial adjustments to temporal variations in flood regime in some Australian rivers. In. K. Richards, (ed.) *River Channels, Environments and Processes*, Blackwell, London.

Warner, R.F. (1987b) The impacts of alternating flood- and drought-dominated regimes on channel morphology at Penrith, New South Wales, Australia. *I.A.H.S. Publ.*, 168, 327-338.

Warner, R.F. and Paterson, K.W. (1987) Bank erosion in the Bellinger Valley, NSW: definition and management. *Aust. Geog. Studies*, 35, 3-14.

Wellman, P. and McDougall, I. (1974) Potassium-Argon ages on the Cainozoic volcanic rocks of New South Wales. *J. Geol. Soc. Aust.*, 21, 247-272.

Wolman, M.G. and Leopold, L.B. (1957) River flood plains: some observations on their formation. *U.S.G.S. Prof. Paper 282-C*.

Wolman, M.G. and Gerson, R. (1978) Relative scales of time and effectiveness of climate in watershed geomorphology. *Earth Surface Process*, 3, 189-208.

Wolman, M.G. and Miller, J.P. (1960) Magnitude and frequency of forces in geomorphic processes. *J. Geol.*, 68, 54-74.

Young, R.W. (1977) Landscape development in the Shoalhaven River catchment of southeastern New South Wales. *Zeit. f. Geomorph.*, 21, 262-283.

Young, R.W. (1983) The tempo of geomorphological change: evidence from Southeastern Australia. *J. Geol.*, 91, 221-230.

11
Geomorphic Effects of Alternating Flood- and Drought-Dominated Regimes on NSW Coastal Rivers

Wayne D. Erskine

School of Geography
University of New South Wales
Kensington, NSW

Robin F. Warner

Department of Geography
University of Sydney
Sydney, NSW

I. INTRODUCTION

Flood-dominated regimes (FDR) refer to time periods of several decades during which there is a marked upward shift of the whole flood frequency curve. A series of floods, particularly large ones, in rapid succession can totally destroy river channels and floodplains (Schumm and Lichty, 1963; Burkham, 1972). If the period of high flood activity persists for any length of time (say >10^1 years) channel recovery and floodplain reformation are prevented (Erskine, 1986a). If the period of high flood activity does not persist the flood-related channel changes are repaired and a new floodplain is constructed. Although recovery periods are highly variable they are inversely related to mean annual rainfall (Wolman and Gerson, 1978). Hickin (1983) coined the term "flood-dominated channel morphology" for flood-induced transient channel form and Warner (1987a; 1987b) modified this term to FDR for periods of high flood activity which have a duration of several decades. Although FDR are defined solely on the basis of flood activity,

Erskine (1986a) and Warner (1987a; 1987b) clearly indicated
that they are also times of large-scale channel changes.
    Drought-dominated regimes (DDR) are periods of several
decades during which there is a marked downward shift of
the whole flood frequency curve from the previous FDR.   They
do not consist solely of drought years but are characterised
by relatively long periods of low flood activity (Warner,
1987a; 1987b).   Channel recovery from the changes effected
during the previous FDR have been recorded in DDR (Warner,
1987a; 1987b).   Both FDR and DDR are more precisely defined
in a subsequent section.
    The purpose of this chapter is to briefly review the
evidence for alternating FDR and DDR on NSW coastal rivers
since European settlement, further document and characterise
these regime shifts and outline their relevance to channel
adjustment and historical river metamorphosis.

II.   TEMPORAL VARIATIONS IN FLOOD REGIME ON NSW COASTAL
      RIVERS

    Alternating FDR and DDR appear to characterise the flood
records of NSW coastal rivers (Fig. 1) since European
settlement (Warner, 1987a).   A growing body of evidence has
demonstrated unequivocally that these alternations are
causally related to secular rainfall changes and not to
land use changes (Pickup, 1976; Riley, 1980; Erskine and
Bell, 1982; Erskine, 1986a).   Correlative changes in rainfall and floods were first recognised on the Hawkesbury-
Nepean River near Sydney (Fig. 1).   This drainage basin
had been settled by the end of the eighteenth century and
the flood record at Windsor is the longest in Australia
(Riley, 1981).   Hall (1927), from an analysis of hydrological
records, concluded that the magnitude and freuqency of floods
at Windsor exhibited identical trends to catchment rainfall.   In particular, the period 1820-1856 was characterised
by few floods and low rainfall, the period 1857-1900, by
many floods and high rainfall, and the period 1901-1925,
by few floods and low rainfall.   It was also demonstrated
that the water supply dams constructed on the upper Nepean
in the early part of this century did not influence
significantly floods at Windsor. Pickup (1976), Riley (1980;
1981) and Warner (1987b), unaware of Hall's (1927) previous
work, also identified similar FDR and DDR.   The last DDR
of Hall continued to about 1948 when another FDR commenced
(Pickup, 1976; Warner, 1987b).   Ironically, Riley (1981)

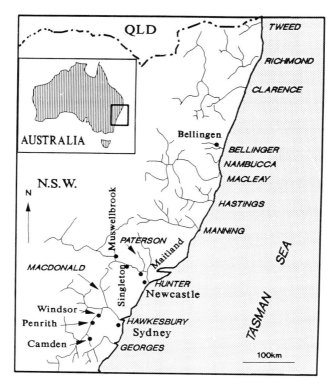

FIGURE 1. NSW coastal rivers north of the Georges
River.

concluded that secular rainfall changes had a greater impact
on the flood regime at Windsor than the construction of a
large water supply dam on the lower Warragamba River in
1960, a result consistent with the earlier findings of
Hall (1927).

Kraus (1955) first recognised that annual rainfall in
NSW was much greater in the latter part of last century than
in the first half of this century. More systematic work
by Gentilli (1971) documented substantial decreases (up to
75 mm) in mean annual rainfall for the period 1911-1940 over
1881-1910 throughout most of coastal NSW. Annual rainfall
increased abruptly in the mid-1940s (Kraus, 1955) and this
increase persisted until at least the late 1970s (Cornish,
1977). Increases in mean annual rainfall exceeded 30% in
some coastal areas (Cornish, 1977; Bell and Erskine, 1981;
Erskine, 1986a; 1986b) and were caused largely by a

significant increase (up to 65%) in summer rainfall (Cornish, 1977; Erskine, 1986a; 1986b). Erskine (1986b) demonstrated that the recent increase in rainfall was accompanied by a statistically significant increase in the number of daily rains of various sizes (rainfall frequency) and that rainfall intensities (24 and 48 hours duration) of storms with return periods of less than 10 years also increased by up to 40% (see also Erskine and Bell, 1982).

Secular changes in mean annual rainfall reflect similar changes in summer rainfall and in rainfall frequency and intensities. Therefore, increases in mean annual rainfall would be expected to cause a contemporaneous increase in flood frequency and vice versa. Although Hall (1927) and Pickup (1976) were the first to recognise this, they did not attempt to show that the trends and inter-relationships were statistically significant. Bell and Erskine (1981) were the first to demonstrate that the post-1946 increase in rainfall produced statistically significant increases in annual runoff and flood frequencies. There are now many examples of an abrupt increase in flood frequency since the mid- to late-1940s on NSW coastal rivers. The best documented cases are for the rivers north of Sydney (Fig. 1), including the Tweed, Richmond, Clarence (Smith and Greenaway, 1982), Bellinger (Warner, 1987a), Macleay (Milne, 1971), Hunter (Erskine and Bell, 1982), Colo, Macdonald (Erskine, 1986a) and Hawkesbury-Nepean Rivers (Warner, 1987b). In the next section, long-term flood records are further examined for evidence of alternating FDR and DDR.

III. FURTHER EVIDENCE OF ALTERNATING FLOOD REGIMES

Six gauging stations on the Hawkesbury-Nepean, Hunter and Bellinger Rivers (Fig. 1) with long records of flood heights were selected for analysis (Table I). Reliable, essentially complete records usually do not extend back beyond 1850 although Windsor and Maitland are exceptions. Hall (1927) and Erskine (1986b) noted a substantial increase in flood incidence after 1857 at Windsor and Maitland, respectively. Therefore, the present work is restricted to the period since 1857 (the records for two stations start after this date). The flood records are partial series of flood heights above a specified threshold (Table I) which generally corresponds to at least bankfull stage. Flood height records must be used for the

present analysis because continuous recording of water
surface elevations and systematic velocity gaugings did not
start in NSW until the 1890s, precluding the use of annual
series of peak instantaneous discharge.

Each flood record was split into the following three
periods: 1857/70-1900, 1901-1946 and 1947-1978/87. The first
period corresponds to FDR, the second to a DDR, and the third
to another FDR. Separate partial series flood frequency
curves were determined for each period, using the
distribution-free plotting formula of Cunnane (1978) and
those for Windsor and Maitland are shown in Fig. 2.
Clearly the flood frequency curves for the DDR exhibit
substantial reductions in flood frequency (ie increase in
recurrence interval for a given flood height) over FDR.
However, while the curves for both FDR at Windsor are
similar, those at Maitland are quite different. Later an
attempt is made to see if these differences between FDR and
DDR as well as between the two FDR are statistically
significant.

Changes in flood-frequency(ie the number of floods in
various height classes) and flood height were tested.   A
chi-square test was used to determine whether flood
frequency is greater during FDR than during DDR (one-tailed
test). Table II clearly demonstrates that the two FDR have
significantly greater flood frequencies than the DDR at all
stations.   These results are similar to Riley's (1980) and
indicate that the change in flood regime involves
significant variations in the number of floods of a given
height. A chi-square test was also used to determine
whether flood frequency differs between the two FDR (two-
tailed test). The results (Table II) show that there is no
significant difference between the two FDR at three stations
but that there is a significant difference at three others.
However, the two stations on the Hawkesbury-Nepean River
are the only ones where the difference is significant at
the 5 percent level or less. Five major water supply dams
as well as other large dams, weirs and farm storages have
been constructed in this basin since 1907. It is suspected
that the difference in flood frequency between the two FDR
on the Hawkesbury-Nepean River has been caused by increasing
storage of surface water in dams and weirs during the present
century.

The nonparametric Mann-Whitney U test has been
recommended by hydrologists to determine whether two
independent samples are drawn from the same population (Kite,
1977). In the present case, the Mann-Whitney U test was
used to determine whether the flood heights for FDR are
significantly greater than those for DDR (one-tailed test).

TABLE I. Partial Series Flood Height Records Analysed for Variations in Flood Regime

Gauging Station and Index No.	Catchment Area (km²)	Period of Record	Threshold Gauge Height (m)	Source of Flood Record
Nepean River at Camden (212900)	1380	1860-1987	9.00	Green (1987)
Hawkesbury River at Windsor (212903)	12770	1857-1978	8.00	Riley (1980)
Hunter River at Muswellbrook (210002)	4220	1857-1978	6.00	Water Resources Commission (1985)
Hunter River at Singleton (210001)	16400	1857-1978	8.00	Records of Department of Water Resources, Moriarty (1870), Bell (1899)
Hunter River at Maitland (210047)	17500	1857-1978	8.23	Erskine (1986b)
Bellinger River at Bellingen (205900)	650	1870-1979	8.00	Department of Public Works (1979)

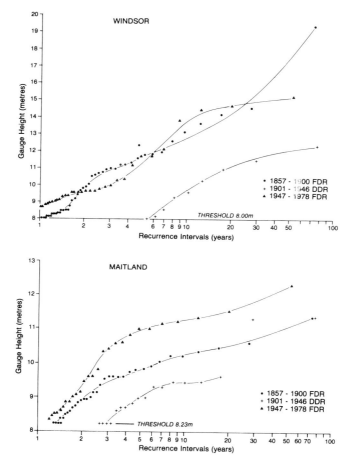

FIGURE 2. Partial series flood frequency curves for FDR and DDR at Windsor and Maitland.

Table II shows that no significant differences were found between both FDR and the single DDR at two stations, that the first FDR had significantly greater flood heights than the DDR at three stations and that the last FDR had significantly greater flood heights than the DDR at one station. Although FDR are characterised by a significant increase in flood frequency over DDR, flood heights do not always increase. The Mann-Whitney U test was again used to test the null hypothesis that flood heights for both FDR belong to the same distribution (two-tailed test). Table II shows that there is no significant difference at three stations and a significant difference at three others. The

TABLE II. Tests of Significance for Differences in Flood Characteristics Between FDR and DDR and Between the Two FDR.

Station	Period of Record	Flood Frequency	Flood Heights
Nepean River at Camden	1860-1900 v 1901-1946[1]	**	**
	1901-1946 v 1947-1987[1]	***	NS
	1860-1900 v 1947-1987[2]	**	****
Hawkesbury River at Windsor	1857-1900 v 1901-1946[1]	****	NS
	1901-1946 v 1947-1978[1]	****	NS
	1857-1900 v 1947-1978[2]	**	NS
Hunter River at Muswellbrook	1857-1900 v 1901-1946[1]	**	*
	1901-1946 v 1947-1978[1]	**	NS
	1857-1900 v 1947-1978[2]	*	*
Hunter River at Singleton	1857-1900 v 1901-1946[1]	**	**
	1901-1946 v 1947-1978[1]	**	NS
	1857-1900 v 1947-1978[2]	NS	NS
Hunter River at Maitland	1857-1900 v 1901-1946[1]	**	NS
	1901-1946 v 1947-1978[1]	***	**
	1857-1900 v 1947-1978[2]	NS	**
Bellinger River at Bellingen	1870-1900 v 1901-1946[1]	**	NS
	1901-1946 v 1947-1979[1]	**	NS
	1870-1900 v 1947-1979[2]	NS	NS

1 - 1 tailed test   2 - 2 tailed test

NS - not significant; * - $p < 0.10$; ** - $p < 0.05$; *** - $p < 0.01$; **** - $p < 0.001$

above results suggest that there is just as significant a difference in flood heights between FDR as there is between FDR and DDR.

FDR are characterised by a significant increase in the number of floods in various height classes (flood frequency) over DDR. These increases in flood frequency may also be accompanied by significant increases in flood heights. Although it is possible to discriminate between FDR and DDR it does not necessarily follow that all FDR are the same. Thus significant differences in flood frequency and flood heights between separate FDR can be expected.

Investigations of changes in flood heights and frequencies over time at gauging stations should also consider changes in the rating curve (ie the stage-discharge relationship) because flood changes may be symptomatic of changes in channel capacity. Many studies of historical changes in flood heights and frequencies have failed to assess whether the changes were caused solely by an alteration of the rating curve for the station (for an example, see Howe *et al*, 1967). Although there is a lack of velocity gauging data for the first FDR at all stations, there are sufficient data for Maitland to determine if the latest FDR has been caused by reduced channel capacity.

Figure 3 is a scatter diagram of gauge height versus discharge for all reliable Department of Water Resources gaugings above 4.57 m since 1913 (the time of the first current meter gauging). No attempt has been made to define individual rating curves for any of the time periods. It is possible, however, to separate two broad groups of points by the line on Figure 3. Gaugings before 1955 generally plot above the line for a discharge less than 2000 m^2/s, whereas those after 1955 plot below. This means that for discharges less than about 2000 m^2/s, the corresponding gauge height since 1955 is less than it was before 1955. Further evidence of this reduction in gauge height for a given discharge at Maitland since 1955 was found by Kazimierczuk (1975) who correlated all corresponding flood peaks between the Maitland and Paterson River at Hinton gauges since 1921. He also found that there were two distinct groups of points with flood heights at Maitland since 1955 being significantly lower for a given flood height at Hinton than those before 1955. This result is important because the Paterson River channel at Hinton has not changed since 1858 (Harrison, 1957). These analyses are consistent and clearly demonstrate that channel capacity at Maitland has increased not decreased since 1955. This proves that the latest FDR is not a product of a rating curve change. In fact, the

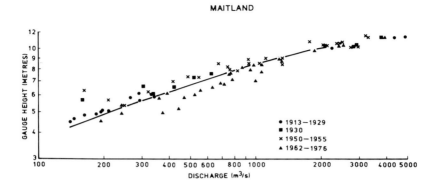

FIGURE 3.   *Changes in the gauge height/discharge relationship between 1913 and 1976 at the Maitland gauge.*

reduced gauge heights since 1955 for a flood peak of less than 10.0 m biases the data against detecting the observed significant increase in flood frequencies and heights.

IV.   GEOMORPHIC EFFECTS OF ALTERNATING FLOOD REGIMES

The geomorphic effects of alternating FDR and DDR on rivers can only be determined from reliable data on channel and floodplain conditions over time.   Such data have been compiled for the Hunter River at Maitland and Nepean River at Penrith (Fig. 1).

A.   *Hunter River at Maitland*

The Maitland study reach is shown in Figure 4 and historical river metamorphosis here was investigated in detail by Erskine (1986b).   The present account is a summary of that work (pp.47-79) which contains much additional information.   Both planform and cross-sectional channel changes are outlined below.

1.   *Planform Changes.*   Although the Hunter River immediately upstream of Maitland has remained laterally stable since European settlement, the downstream section between Maitland and Morpeth, which, unlike the upstream section, is essentially unconfined by bedrock, has under-gone dramatic change.   Maps compiled by Mitchell (1823) and

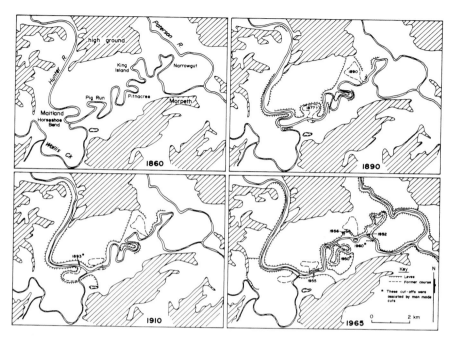

FIGURE 4. History of cutoffs and artifical levee con-struction on the Hunter River between Oakhampton and Morpeth (after Nittim, 1966).

Moriarty (1870), among others, show that the Hunter River was very sinuous early last century and this is illustrated by the 1860 planform of Figure 4. In 1870 sinuosity was 3.84 (Table III) which far exceeds any of Schumm's (1963) and Begin's (1981) data. It is also greater than the sinuosity of the Greenville reach of the Mississippi River before the infamous cutoffs of the 1930s (Schumm *et al*, 1972). Historical sources listed by Erskine (1986b) indicate that sinuosity fluctuated little from 3.84 between 1823 and 1870. Three loops were abandoned during the FDR of late last century, resulting in a marked straightening of the channel (Fig. 4; Table III). Horseshoe Bend was abandoned following the excavation of a diversion canal through the neck of the loop. The other two cutoffs, King Island and Pig Run, were abandoned by natural geomorphic processes. These three cutoffs shortened channel length between Maitland and Morpeth by 8 km over the 1870 value, thus reducing sinuosity to 2.66 (Table III). Another five cutoffs occurred between 1949 and 1960 at the beginning of the second FDR (Fig. 4). Excavations were carried out at two sites

TABLE III.   *Changes in sinuosity (ie ratio of channel length to valley length) since 1870 for Hunter River between Maitland and Morpeth.   Valley length is 6.8 km.*

Year	Channel Length (km)	Sinuosity (km/km)	Channel Pattern (after Schumm, 1963)
1870	26.1	3.84	Tortuous
1899	18.1	2.66	Tortuous
1952	9.7	1.43	Regular
1964	9.4	1.38	Transitional

to either promote a cutoff or hasten the development of a pre-existing cutoff channel (Fig. 4).   The post-1949 cut-offs further shortened stream length between Maitland and Morpeth by 8.7 km, thus reducing sinuosity to 1.38 (Table III).

Since 1870 there have been eight cutoffs in the reach between Maitland and Morpeth which have reduced channel length by 16.7 km to 9.4 km with a consequent decrease in sinuosity from 3.84 to 1.38 (Fig. 5).   As a result, river planform has changed from tortuous to transitional in terms of Schumm's (1963) classification or from meandering to straight in terms of Leopold and Wolman's (1957) channel patterns.   These cutoffs only occurred during FDR although excavations were carried out in some cases to hasten cut-off development.   It is interesting to note that in 1870, Moriarty (1870), as part of a flood mitigation scheme for Maitland, proposed a number of channel diversions between Maitland and Morpeth to reduce channel length to 9.6 km. By 1964 these diversions had occurred largely by natural processes.

2.   *Cross-Sectional Changes.*   Between 1812 and at least 1836 the banks and floodplain of the Hunter River at Maitland were densely vegetated by forest which had been largely cleared by 1870 (Hunter River Floods Commission (HRFC), 1870b).   This clearing, combined with the onset of the FDR in 1857 resulted in extensive bank erosion (HRFC, 1870a; 1870b; Moriarty, 1870) which was still active in 1899 (Bell, 1899).

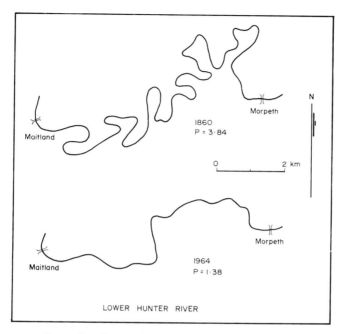

FIGURE 5. Planform changes in the lower Hunter between Maitland and Morpeth since European settlement (modified after Nittim, 1966). P denotes sinuosity.

Flooding and bank erosion were perceived to be such a major problem during the FDR of late last century that a Royal Commission was appointed to investigate the matter and propose possible solutions (HRFC, 1870a; 1870b). Although some of the evidence taken by the Commission refers to significant aggradation at Maitland by 1870, the Commission did not support this view. In August 1870, the Commission had resurveyed eight cross sections between Oakhampton and Morpeth which were originally surveyed in 1857. A comparison of these surveys showed that:

> "...the channel has not diminished in depth or magnitude but, on the contrary, that it has sensibly increased" (HRFC, 1870b, p.26).

In fact, repeated surveys at the Maitland gauge between 1865 and 1965 reveal that the channel was wider in 1865 than at any time since (Erskine, 1986b).

Aggradation had become a major problem by 1947 when Huddleston *et al* (1948) drew attention to massive sedimentation in the channel at Maitland. Figure 6 shows

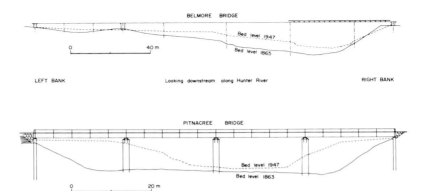

FIGURE 6.   Cross-section changes at Belmore and Pitnacree bridges (after Huddleston et al, 1948).   Sections are plotted without vertical exaggeration.

their comparison of cross sections surveyed in 1865 and 1863 at Belmore and Pitnacree bridges, respectively, with those resurveyed in 1947.   Huddleston *et al* (1948) believed that this aggradation had affected the whole river between Maitland and Hexham and had increased flood heights (this is refuted in the previous section).   Subsequently, Bernard (1950) suggested that this aggradation was restricted to the period 1930-1949 (Fig. 7).   Erskine (1986b), using surveys undertaken by the Department of Water Resources showed that bed aggradation only occurred between 1930 and 1950, although bank sedimentation and overbank deposition had been active since 1913 (Fig. 7).   The aggraded bed material was sand and it buried the previous gravel riffles and sand and gravel pools.   Channel widening, bank erosion and cutoffs were characteristic of the FDR between 1857 and 1900 whereas the following DDR was characterised by substantial channel and floodplain deposition.

At the very time that Huddleston *et al* (1948) and Bernard (1950) were drawing attention to bed aggradation, river response changed.   Erskine (1986b) demonstrated that the latest FDR initiated a period of slight channel widening and degradation.   Holmes and Loughran (1976) used paired vertical air photographs of 23 October 1946 and 5 March 1955 to illustrate a doubling of width over this period. However, the substantial difference in water surface elevations between these two dates (ie 5.48 m) exaggerates the increase in width.

Artificial levees (Fig. 4) have been frequently invoked

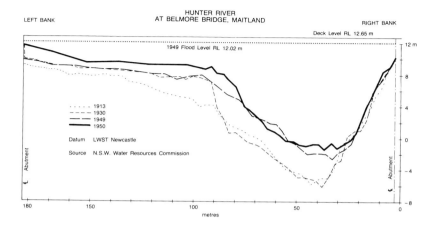

FIGURE 7. Channel changes at the Maitland gauge between 1913 and 1950.

as at least partly causing the channel changes outlined above (Huddleston *et al*, 1948; Harrison, 1957). The mechanisms linking levees to channel changes supposedly involve an increase in channel capacity and flood heights by blocking floodways and floodplain storages, thus confining larger flows to the channel. The larger channel capacity means an increase in bankfull discharge over the natural conditions. Therefore, meander wavelength should increase resulting in a decrease in sinuosity. However, although the predicted planform changes have occurred, Erskine (1986b) has shown that there is no temporal correlation between levee construction and increased discharge or flood height. The main phase of levee construction was between 1870 and 1895 (Fig. 4) and was followed immediately by a DDR during which flood heights did not increase. The greatest increases in flood heights occurred after 1948 when the levee system was extensively damaged by floods and was redesigned to allow overbank flow (Harrison, 1957). This was also a time when water surface elevation for a given discharge decreased, not increased.

B. Nepean River at Penrith

The Penrith study reach (Fig. 8) is located in the Penrith Weir pool and covers the transition of the Nepean River from a confined bedrock gorge to an alluvial plain. The following five zones have been mapped in the study reach (Fig. 8):

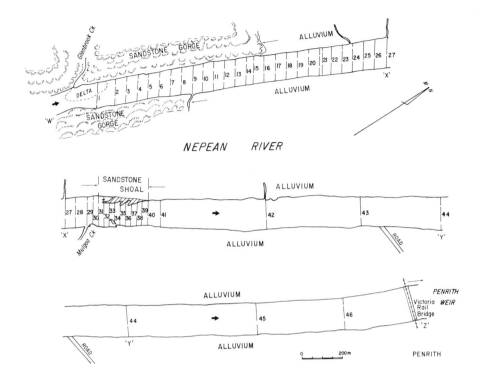

FIGURE 8. Nepean study reach showing locations of 46 cross-sections.

(1) the Main Gorge, where Triassic sandstone frequently forms both banks. The Glenbrook Creek delta is located at the upstream end of this zone and is composed of sandstone boulders eroded from Glenbrook Creek during a large flood in May 1943;

(2) the Lower Gorge where sandstone is only found on the left bank;

(3) the Upper Alluvium where both banks are composed of unconsolidated fluviatile sediments;

(4) the Shoal where a wide sandstone platform outcrops over 274 m of the left bank; and

(5) the Lower Alluvium where both banks are again composed of unconsolidated fluviatile sediments.

Channel capacity is exceptionally large in the alluvial reaches with the 1:100 year flood not producing overbank flow, only backwater flooding on tributaries. The total basin area at Penrith Weir is 11,008 km^2, of which 8450 km^2 is drained by the Warragamba River. Warragamba Dam, located 4 km upstream of the confluence with the Nepean River, was closed in 1960, effectively stopping the supply of sand and gravel to the Nepean River.

The Nepean immediately above Penrith Weir has dimensional records going back to 1862 (start of a FDR) when the Victoria Railway Bridge was designed (Fig. 8). The channel then was relatively narrow and deep. After the 1900 flood, the second largest on record, the channel here had increased in width by 30 m. At this time 46 cross sections were surveyed in the 5 km reach above the bridge, presumably to determine the capacity of the weir which was built in 1909. These 46 cross sections were relocated by trigonometry and resurveyed in 1982/3. The results revealed an increase in capacity of 660,000 m^3 at weir-crest level for the surveyed reach. Four cross sections surveyed in 1938 showed a narrower and deeper channel than in 1900. These latter sections were resurveyed in 1968 by which time the channel had degraded well above the weir but aggraded near to it. The effects of sediment starvation due to the closure of Warragamba Dam and the formation of the Glenbrook Delta were evident, as was the bed-material trapping role of the down-stream weir. These trends continued when the above 4 plus 2 extra cross sections added in 1968 were resurveyed in 1984. Much of the morphological evidence for channel response to both regime shifts and regulatory structures has been summarised by Warner (1987b).

Width trends from 1900 to 1982/3 are shown in Table IV. The 1900 and 1982/3 data were obtained from the afore-mentioned surveys whereas the 1949 and 1970 data were obtained from vertical air photographs. During the DDR (1901-1948) the 46 sections decreased in width by an average of 23 m, with a maximum of 55 m being recorded in the Shoal where the sandstone ledges acted as nuclei for sedimentation. In the following FDR the channel increased in width by 13 m, with the largest change again in the Shoal (35 m). As expected changes in the Main and Lower Gorges were small. In the alluvial sections away from the Shoal, width changed from 165 m in 1900 to 145 m in 1949 to 156 m in 1970 to 157 m in 1982/3. Thus width decreased by 20 m during the DDR and increased by 12 m during the succeeding FDR, with dam impacts being superimposed on regime shifts since 1960. Most of the increase occurred between 1949 and 1970, so adjustments have slowed down appreciably.

TABLE IV. *Summary of Width Changes at Weir Crest Level for 46 Channel Cross-Sections on the Nepean River at Penrith.*

River Zone	1900	Mean Channel Width (m) 1949	1970	1982/3	Change in Width (m) 1900-49	Change in Width (m) 1949-82/3	Cross-Section Numbers
Main Gorge	149	142	141	144	-7	2	1-8
Lower Gorge	159	155	158	157	-4	2	9-16
Upper Alluvium	160	145	155	154	-15	9	17-30
Shoal	163	108	140	143	-55	35	30-40
Lower Alluvium	175	146	158	165	-29	19	41-46
Total	161	138	150	151	-23	13	46

V. DISCUSSION AND CONCLUSIONS

The coastal rivers of NSW, particularly those north of Sydney, exhibit periods of alternating high and low flood activity. FDR are characterised by significantly more and, in some cases, higher floods than the intervening DDR. The duration of these two types of flood regime varies between 30 and 50 years and is controlled by secular rainfall changes. These rainfall changes vary, in turn, with the strength of the Southern Oscillation Index (a measure of the strength of an east-west pressure difference between the eastern tropical Pacific and the area around Darwin) and the latitudinal path of the sub-tropical high pressure belt. DDR have been recorded early last century (1821-1856) and early this century (1901-1946) whereas FDR have been recorded in the intervening periods, 1857-1900 and 1947-1978. The exact timing of the onset and termination of these alternating flood regimes probably varies slightly from basin to basin. Hydrological changes induced by land use changes and flow regulation are superimposed on these natural regime shifts.

Channel response to alternating flood regimes mainly involves widening during FDR with recovery by contraction during DDR. The magnitude of widening during FDR is functionally related to bank resistance with the greatest increases occurring on rivers or on sections of a river with the most erodible boundary sediments (usually sands) and least vegetation. Similarly subsequent contraction is greatest where there is a large supply of bed material and wash load, or where there is a suitable nucleus for sedimentation. Other forms of channel response to FDR have included channel straightening in cutoffs, bed aggradation, where channel widening is substantial and bed degradation, where channel widening is minor. Rather surprisingly, there is little evidence at present of increased rates of lateral migration during FDR. The geometry of NSW coastal rivers appears to vary cyclically in response to changes in flood regime. Therefore, most rivers are continually changing and do not become adjusted to some fixed discharge level.

ACKNOWLEDGEMENTS

The authors are grateful to Drs Mike Melville and Fred Bell for their help in discussions on this essay.

REFERENCES

Begin, Z.B. (1981) The relationship between flow-shear
    stress and stream pattern. *J. Hydrol.*, 52, 307-319.
Bell, F.C. and Erskine, W.D. (1981) Effects of recent
    increases in rainfall on floods and runoff in the upper
    Hunter Valley. *Search*, 12, 82-83.
Bell, N. (1899) *Flood Prevention in the Hunter River*,
    NSW Govt. Printer, Sydney.
Bernard, R.L. (1950) *Report on Flood in Hunter River, June
    1949*, Water Conservation and Irrigation Commission,
    Sydney.
Burkham, D.E. (1972) Channel changes in the Gila River in
    Safford Valley, Arizona 1846-1970. *U.S.G.S. Prof. Paper
    655-G*.
Cornish, P.M. (1977) Changes in seasonal and annual
    rainfall in New South Wales. *Search*, 8, 38-40.
Cunnane, C. (1978) Unbiased plotting positions - a
    review. *J. Hydrol.*, 37, 205-222.
Department of Public Works (1979) *Bellinger River Flood
    History 1843-1979*, Dept. of Public Works, Sydney.
Erskine, W.D. (1986a) River metamorphosis and environmental
    change in the MacDonald Valley, New South Wales since
    1949. *Aust. Geog. Studies*, 24, 88-107.
Erskine, W.D. (1986b) River metamorphosis and environmental
    change in the Hunter Valley, New South Wales. Unpubl.
    PhD Thesis, Univ. NSW.
Erskine, W.D. and Bell, F.C. (1982) Rainfall floods and
    river channel changes in the upper Hunter. *Aust. Geog.
    Studies*, 20, 183-196.
Gentilli, J. (1971) Climatic fluctuations. In *Climates
    of Australia and New Zealand, World Survey of Climatology*,
    Vol. 13, Elsevier, Amsterdam, 189-211.
Green D.W. (1987) Geomorphic impacts of an extractive
    industry on the Nepean River at Cobbitty. Unpubl. BSc
    Thesis, Univ. NSW.
Hall, L.D. (1927) The physiographic and climatic factors
    controlling the flooding of the Hawkesbury River at
    Windsor. *Proc. Linn. Soc. NSW*, 52, 133-152.
Harrison, E.W. (1957) Flood mitigation on the lower Hunter
    River, NSW. *J. Inst. Engrs. Aust.*, 29, 321-331.
Hickin, E.J. (1983) River channel changes: retrospect and
    prospect. *Int. Ass. Sediment. Spec. Publ.*, 6, 61-83.
Holmes, J.H. and Loughran, R. (1976) Man's impact on a
    river system: the Hunter Valley. In J.H. Holmes (ed.)
    Man and the Environment: Regional Perspectives, Longman,
    Hawthorn, 96-114.

Howe, G.M., Slaymaker, H.O. and Harding, D.M. (1967) Some aspects of the flood hydrology of the upper catchments of the Severn and Wye. *Trans. Inst. Brit. Geog.*, 41, 33-58.

Huddleston, G., Green, E.O.K. and Kaleski, L.G. (1948) *Report of Hunter River Flood Mitigation Committee*, NSW Govt. Printer, Sydney.

Hunter River Floods Commission (1870a) *Progress Report of Commission Appointed to Enquire into and Report Respecting Floods in the Hunter*, NSW Govt. Printer, Sydney.

Hunter River Floods Commission (1870b) *Report of Commission Appointed to Enquire into and Report Respecting Floods in the District of the Hunter River*, NSW Govt. Printer, Sydney.

Kazimierczuk, E. (1975) *Flood Mitigation in the Lower Hunter Valley, Vol. 1, Flood Mitigation Works*, Dept. of Public Works, Sydney.

Kite, G.W. (1977) *Frequency and Risk Analyses in Hydrology*, Water Res. Publ., Fort Collins.

Kraus, E.B. (1955) Secular changes of east-coast rainfall regimes. *Q. J. Roy. Met. Soc.*, 81, 430-439.

Leopold, L.B. and Wolman, M.G. (1957) River channel patterns: braided meandering and straight. *U.S.G.S. Prof. Paper 282-B*.

Milne, A.K. (1971) Underfit analysis of the lower Macleay River basin: problems of the application of a model. *Geog. Educ.*, 239-252.

Mitchell, T.L. (1823) *Map of the Surveyed Part of Hunters River*, State Archives Authority of NSW, Map No. 6280.

Moriarty, E.O. (1870) Report on the prevention of floods in the Hunter. In Hunter River Floods Commission, *Report of Commission Appointed to Enquire into and Report Respecting floods in the District of the Hunter River*, NSW Govt. Printer, Sydney, 95-112.

Nittim, R. (1966) Flood studies on the lower Hunter River, New South Wales. Unpubl. M.Tech. Thesis, Univ. NSW.

Pickup, G. (1976) Geomorphic effects of changes in river runoff, Cumberland basin NSW. *Aust. Geog.*, 13, 188-193.

Riley, S.J. (1980) Aspects of the flood record at Windsor, *Proc. 16th Conf. Inst. Aust. Geog.*, Newcastle, 325-340.

Riley, S.J. (1981) The relative influence of dams and secular climatic change on downstream flooding, Australia. *Water Res. Bull.*, 17, 361-366.

Schumm, S.A. (1963) Sinuosity of alluvial rivers on the Great
   Plains. *Geol. Soc. Amer. Bull.*, 74, 1089-1100.
Schumm, S.A., Khan, H.R., Winkley, B.R. and Robbins, L.G.
   (1972) Variability of river patterns. *Nature*, 237,
   75-76.
Schumm, S.A. and Lichty, R.W. (1963) Channel widening and
   flood-plain construction along Cimarron River in
   southwestern Kansas. *U.S.G.S. Prof. Paper 352-D.*
Smith, D.I. and Greenaway, M.A. (1982) Flood probabilities
   and urban flood damage in coastal northern New South
   Wales. *Search*, 13, 312-314.
Warner, R.F. (1987a) Spatial adjustments to temporal
   variations in flood regime in some Australian rivers.
   In K. Richards (ed.) *River Channels: Environment and
   Process*, Blackwell, Oxford, 14-40.
Warner, R.F. (1987b) The impacts of alternating flood- and
   drought-dominated regimes on channel morphology at
   Penrith, New South Wales, Australia. *I.A.H.S. Publ. No.
   168*, 327-338.
Water Resources Commission (1985) *Muswellbrook Flood Study
   Report*, 1st Edition, Water Res. Commission, North
   Sydney.
Wolman, M.G. and Gerson, R. (1978) Relative scales of time
   and effectiveness of climate in watershed geomorphology.
   Earth Surf. Proc., 3, 189-208.

12

Secular Change in the Annual Flows of Streams in the NSW Section of the Murray-Darling Basin

S.J. Riley

School of Earth Sciences
Macquarie University
NSW, Australia

I. INTRODUCTION

The Murray-Darling basin is the largest exorheic drainage basin in Australia (1,062,530 km^2; Brown, 1983). It plays a key role in the Australian economy. A significant proportion of the country's irrigation, cultivation and pastoral activities is undertaken within it (Pigram, 1986). The insidious spread of salt throughout the alluvial soils, particularly noticeable in the lower Murray and Murrumbidgee in the last 30 years, has led to the loss of significant areas of previously highly productive agricultural land (Cole and McCloud, 1981). Successful landuse management in the future will depend on a thorough understanding of the catchment's hydrology.

Essential to an understanding of the nature of the catchment's hydrological regime is a consideration of the secular climatic changes for periods of up to 50 years that have occurred over the last century. The operation of dams, canals and irrigation areas is influenced by associated changes in river flow and water supply (Lettenmaier and Burgess, 1978). Channel geometry and sediment loads also respond to changes in flow regime. The mobilisation of salt, which has been so noticeable in recent years (River Murray Commission, 1970; Collett, 1978; Pigram, 1986) may also be as much attributable to recent secular climatic change as to human activities.

Hydrometeorological changes cause variations in the flow

regime, which include changes in annual floods, flood frequency, annual flows, peak discharges and flow duration. This essay is concerned with the nature of secular change in annual runoff of the NSW streams of the Murray-Darling catchment. The climatic factors that may be responsible for this change are also examined.

Initially the nature of secular climatic change in NSW is reviewed. It is shown that the rainfall regime of the period 1900 to 1946 is different from the preceeding and succeeding periods. Then the characteristics of the NSW Murray-Darling basin as they relate to secular change in annual flows are discussed. Details of the analysis are given in the following section prior to presentation of the results. Finally, the spatial, magnitude and geomorphic significance of the secular changes in annual flows is discussed.

II.   SECULAR CLIMATIC CHANGE

Early studies by Kraus (1954) and Gentilli (1971) and more recent studies have shown that there have been significant secular changes in the rainfall of Eastern Australia (Pittock, 1975; Australian Academy of Science, 1976; Cornish, 1977). The studies show that the period prior to 1900 and that from the mid-1940s to the present have been much wetter than the period from 1900 to 1945. Less is known of the rainfall pattern of the earliest period because of problems associated with record quality and paucity. Gentilli (1971) only extended his study back to 1881 and the most reliable rainfall records extend from 1900 to the present. As the same problem of quality also exists for the flow records, this essay concentrates on the secular changes in the 20th century.

Rainfall records show an increase in the annual totals and an increase in the incidence of storms for the whole of Eastern Australia in the 1946-1985 period from that of 1900-1945. Cornish (1977) showed that annual rainfall in New South Wales increased by up to 30% for the period between 1946 and 1973, relative to the preceeding years of record. The greatest increase was in the central region of NSW, in an area that extends between Yass and Sydney to Cobar and Walgett. These changes were in excess of 10 to 20%. The lowest increase was recorded along the NSW-Victorian border, some areas with no change. The majority of the increase was accounted for by an increase in summer rainfall, which was approximately 60% around Sydney and greater than 30% in the central region defined above. Except for a region around Balranald and Hay, the summer change was small along the Victorian border, and the upper Murray even showed a decrease.

Pittock (1978) attributed the changes in eastern Australian rainfall to the Southern Oscillation and the mean latitude of the high-pressure belt. He noted that "the change which occurred in rainfall over southeast Australia between the two roughly 30-year periods before and after the mid 1940s corresponds rather well to the pattern associated with year-to-year variations in the Southern Oscillation" (p.175).

Runoff records also show significant increases in total flow, flood magnitude and flood frequency between the two periods 1900-1945 and 1945 to the present (Bell and Erskine, 1981; Erskine and Bell, 1982). The increases have been so large that even major engineering structures, such as Warragamba dam, have failed to mask the change (Riley, 1981).

The hydrological and consequent morphological impacts of the climatic changes are not perfectly understood and they may not have fully worked themselves out in the catchments; lag times are still largely unknown. It is thought that the recent 'catastrophic' changes noted in the Hunter, MacDonald and Manning Rivers are also related to an increase in flood frequency and flood magnitude (Erskine and Bell, 1982; Henry, 1977; Nanson, 1986; Erskine and Warner, this volume). Increased gullying and mobilisation of channel sediment have been observed in a number of catchments throughout Eastern Australia, particularly eastward-flowing streams (Pickup, 1976) and there is general evidence that these streams are far from stable (Erskine, 1986). If, as argued by several authors (Pickup and Warner, 1976; Neller, 1980), river channel geometry is adjusted to medium and low frequency events, then the increased frequency and magnitude of flooding of the last three decades would have contributed to morphologic change. However, some care needs to be exercised in interpreting the information on 'catastrophic' channel changes in terms of secular climatic change. One reason why these channel changes are being noticed may be better surveillance of river systems.

The economic impacts of these changes are enormous. Increased flood frequency and magnitudes have increased the demands on emergency services. Many dams whose spillways were designed from data collected in the earlier 1900-1945 period have been shown to be underdesigned (NSW Dams Safety Committee, 1985). There are problems associated with

the stability of structures within rivers. Additionally,
complications are evident in assessing the relative impact
of various human activities, such as sand and gravel mining
of channels and natural changes. Whilst a considerable body
of information has been gained concerning the 20th century
secular changes in flow regime of the east-flowing rivers,
less is known about the changes in the west-flowing
streams.

III. MURRAY-DARLING CATCHMENT

The Murray-Darling catchment within NSW has its headwaters in the highland areas of eastern and south-eastern
Australia (Fig. 1). There are major structural controls
on the drainage pattern (Hills, 1972, p.457; Ollier, 1978)
even though much of the catchment is dominated by alluvial
plains over which highly sinuous streams have gradients less
than 0.001, and commonly less than 0.0001. The headwater
rivers are confined in well defined valleys which increase
in relief towards the east. The lower reaches of the rivers
are dominated by wash load (Woodyer, 1978), the upper by
suspended and bedload.

Floods are generated in the east and north east, whilst
in the west and south west of the catchment, the occasional
heavy storms fail to produce major floods because low
gradient plains and depression stores do not facilitate rapid
surface runoff. Average annual rainfalls are as high as
2400 mm in the east and as low as 200 mm in the west. The
majority of the catchment has an average annual rainfall
less than 500 mm. Stream flow is highly variable (Dept.
of Nat. Develop., 1978) particularly in the western two-
thirds of the catchment. Further details on the sub-catchments are given elsewhere (WC & IC, 1968-75; W.R.C. 1982-84).

Although water conservation and irrigation began in the
last century, it is in the period since the late 1940s that
the NSW section of the Murray-Darling basin has been much
influenced by dam construction. The majority of the irrigated
areas are in the downstream, western plains sections of the
streams (Pigram, 1966, p.164) and approximately 4500 GL of
water are used annual from rivers and dams. The Murrumbidgee
and Murray irrigation areas are the largest users of water.

The majority of dams were constructed after 1940 and are
located in the east (WC & IC, 1971; Crabb and Riley, 1987).
Most are now used for water storage rather than flood control.
Numerous weirs were constructed along the streams in the

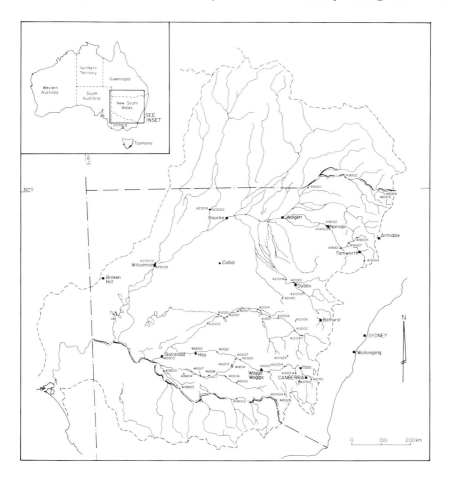

*FIGURE 1.   Murray Darling Basin showing the location
of gauging stations.   Station numbers are after the
Australian Water Research Council (1984).   Stations are
listed in Table I.*

late 1800's, but they have a minimum influence on stream
flow.

Baker and Wright (1978) suggest that irrigation has
reduced flow in the Murray, and this is probably true for
all the major streams with irrigated areas.   The impact of
irrigation and dam construction should be a decrease in
annual flow volumes and variability of flows (Petts and
Lewin, 1979; Goldsmith and Hilgard, 1984).

IV.   METHODOLOGY

Annual flow volume records of the NSW Department of
Water Resources for the Murray-Darling Basin are the basic
data set for this study. Gauging stations were selected
if they had a near complete record for the period 1921 to
1961, considered to be the minimum period needed to assess
adequately regimes before and after the mid 1940s. The
record for the period 1901 to 1986 was extracted, if
availabile.

Whilst there are a number of gauging stations in the
Murray-Darling catchment (AWRC, 1984) only 52 stations had
suitable record (Table I, Fig. 1). Table I records the
AWRC number of the stations, their name and catchment area.
A number of distributary or effluent streams are included
and for these no area is recorded. The period of record
(first to last record) within the period 1901 to 1986 is
noted, as is the number of years with complete records. A
number of stations had records that extended back to 1901
and most had records up to 1984-85, but all stations, except
the Darling River at Wilcannia, had some years of missing
records (Table I).

A.   *Missing Records*

In order to maximise the record period and to reduce
missing values of annual flows, correlation analysis was
used among the stations. The technique of estimating
missing values of annual flow was as follows. Regression
relations were developed between all stations in a major
sub-catchment of the Murray Darling Basin (as defined by the
WRC, 1984)· In the sub-catchment, the station with the
longest, most complete record was taken as the base station.
Missing annual flow values for the base station were
estimated by seeking the highest correlation between the
base station and the other stations in the sub-catchment
and by using the regression relation to predict the missing
value. The station with the next highest correlation
with the base station was used if the first chosen station,
used as the independent variable, also had missing values
for the particular year. The procedure was then repeated
for the station in the sub-catchment with the next least
missing years of annual flows in the 1901-1986 period, and
so on until all the stations in a sub-catchment had missing
annual flows estimated. This statistical procedure for

TABLE I. Station Numbers, Names, Catchment Areas and Years of Record.

Station number	Station name	Catchment area (km2)	Years of record (1901-86)	Years of complete record
409002	MURRAY RIVER AT COROWA ROAD BRIDGE	18700	69	67
409005	MURRAY RIVER AT BARHAM	43300	80	69
409013	WAKOOL RIVER AT STONEY CROSSING	EFFLUENT	64	57
409014	EDWARD RIVER AT MOULAMEIN	EFFLUENT	66	60
410001	MURRUMBIDGEE RIVER AT WAGGA	27700	84	83
410002	MURRUMBIDGEE RIVER AT HAY	56700	82	79
410003	MURRUMBIDGEE RIVER AT BALRANALD	165100	73	72
410004	MURRUMBIDGEE RIVER AT GUNDAGAI	21100	84	83
410005	MURRUMBIDGEE RIVER AT NARRANDERA	34100	85	83
410007	YANCO CREEK AT OFFTAKE	EFFLUENT	83	78
410009	JOUNAMA CREEK AT TALBINGO	134	60	55
410012	BILLABONG CREEK AT COCKETGEDONG	4661	72	67
410014	COLOMBO CREEK AT MORUNDAH	EFFLUENT	66	63
410015	VANKO CREEK AT MORUNDAH (BINGEGONG BRIDGE)	EFFLUENT	67	57
410016	BILLABONG CREEK AT JERILDERIE	EFFLUENT	67	51
410017	BILLABONG CREEK AT CONARGO (PUCKAWIDGEE)	EFFLUENT	68	60
410018	YANCO CREEK AT MOONBRIA	EFFLUENT	50	28
410021	MURRUMBIDGEE RIVER AT DARLINGTON POINT	38800	69	62
410024	GOODRADIGBEE RIVER AT WEE JASPER (KASHMIR)	1160	68	63
410025	JUGIONG CREEK AT JUGIONG	2120	64	48
410029	BUDDONG CREEK AT BUDDONG FALLS (BUDDONG WEIR)	28	51	46
410700	COTTER RIVER AT KIOSK	482	71	66
410701	QUEANBEYAN RIVER AT GOOGONG	872	64	62
410702	MOLONGLO RIVER AT CORKHILLS	1850	46	45
412002	LACHLAN RIVER AT COWRA	11000	83	82
412004	LACHLAN RIVER AT FORBES (IRON BRIDGE)	19000	85	82
412005	LACHLAN RIVER AT BOOLIGAL WEIR	55900	79	74
412009	BELUBULA RIVER AT CANOWINDRA	2170	74	70
412011	LACHLAN RIVER AT LAKE CARGELLIGO WEIR	45800	76	71
412012	WILLANDRA CREEK AT ROAD BRIDGE	EFFLUENT	79	63
412014	GOOBANG CREEK AT CONDOBOLIN	EFFLUENT	70	68
412017	BUMBUGGAN CREEK AT OFFTAKE	EFFLUENT	54	49
412020	WILLANDRA CREEK AT TOCABIL	EFFLUENT	48	46
412067	LACHLAN RIVER AT WYANGALA	8280	66	63
416001	BARWON RIVER AT MUNGINDI	44000	85	82
416002	MACINTYRE RIVER AT BOGGABILLA	22500	82	78
416003	TENTERFIELD CREEK AT CLIFTON	569	60	51
416004	MOLE RIVER AT TRENAYR	1600	50	45
419001	NAMOI RIVER AT GUNNEDAH	17000	81	76
419002	NAMOI RIVER AT NARRABRI	25100	82	79
419004	PEEL RIVER AT BOWLING ALLEY POINT	310	64	51
419005	NAMOI RIVER AT NORTH CUERINDI	2510	71	61
419006	PEEL RIVER AT CARROL GAP	4660	62	59
419007	NAMOI RIVER AT KEEPIT NO2	5690	60	55
421001	MACQUARIE RIVER AT DUBBO (RAILWAY BRIDGE)	19600	83	71
421003	MACQUARIE RIVER AT WELLINGTON	14100	67	63
421006	MACQUARIE RIVER AT NARROMINE	26100	78	74
421010	BOGAN RIVER AT PEAK HILL	543	43	42
423001	WARREGO RIVER AT FORDS BRIDGE	60600	62	55
423002	WARREGO RIVER AT FORDS BRIDGE BYWASH	60600	62	60
425002	DARLING RIVER AT WILCANNIA (TOTAL FLOW, INCLUDES TALYAWALKA)	569000	79	79
425008	DARLING RIVER AT WILCANNIA (MAIN CHANNEL)	569000	73	72

(see text for explanation of columns)

estimating missing annual flows resulted in the majority
of stations having complete records for the period 1901 to
1984.

B.   Cumulative Deviation from the Mean

The corrected set of annual flows for each gauging
station was then examined for changes over the period 1901
to 1986 and for stationarity using a variety of techniques.
Variation of flow over time is shown using graphs of
the cumulative deviation from the mean. To compare stations
with different mean annual flows it was necessary to
standardise them for each station to a mean annual flow of
1.0 by dividing each annual flow by the mean annual flow.
Cumulative deviations are calculated by accumulating the
annual flow for each successive year and subtracting 1.0
at each step. The resulting time series of cumulative
deviation from the mean shows those periods when flows were
less than mean (deviations decrease in value over time) and
periods when flows were greater than the mean (deviations
increase in value). The five-year running mean of annual
flows was examined for each station. The trends were similar
to those produced by the cumulative deviations from the
mean, so the results are not presented.

V.   RESULTS

A.   By Basins

1.   Upper Barwon (Basin 416).   In the Upper Barwon basin,
there are well defined troughs in the cumulative deviation
from the mean (Fig. 2), with the largest negative cumulative
deviation at Boggabilla (416002) and the smallest at
Clifton (416003), where recovery back to zero deviation
only took 3 years. The same sequence of wet and dry years
is present for all four stations. The wetter period appears
to begin in 1946. The 1914-1930 and 1954-1985 periods had
near average flows at Clifton. For the others, there were
below average flows between 1955 and 1970 and above
average flows for 1975 and 1985. The difference in the
response at Clifton on Tenterfield Creek may be related to
catchment geometry and soil and rock materials in the catchments.

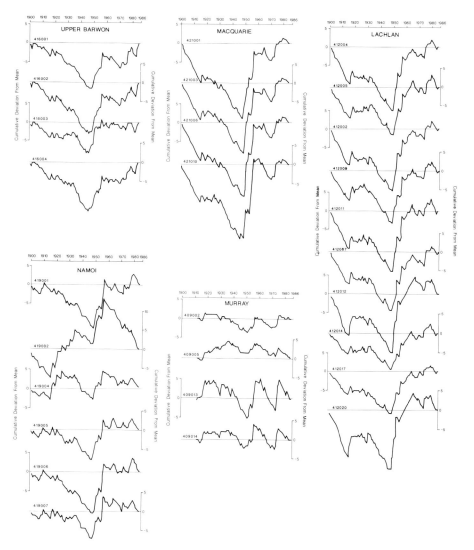

FIGURE 2.   Cumulative deviations from the mean for
stations listed in Table I.   All annual flows have been
standardised to the mean annual flow for the total period
of record.

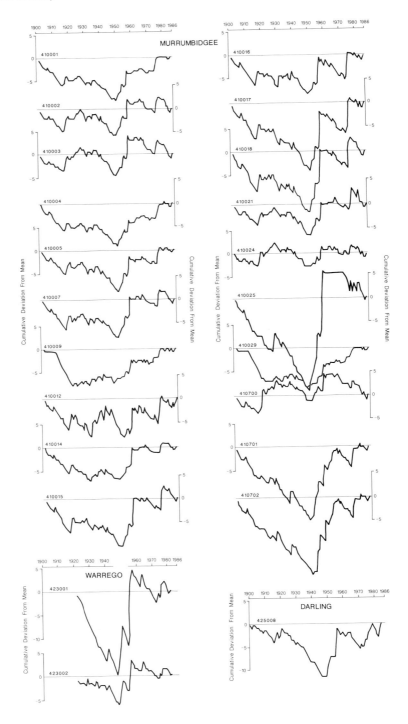

2. *Namoi (Catchment 419)*. Considerable variations in cumulative deviations from the mean are found for the six stations in the Namoi (Fig. 2). Five show minimum departure in the mid 1940s but at Narrabri it is in the mid 1910s. Up to 1906 all show below average flow but this has recovered by 1910 for most. After the dry period of the mid 1940s, all show near maximum departures by the mid 1950s. Thereafter fluctuations are common but not long lasting except at Narrabri, which has shown a decrease throughout the period 1955-1985. The gauge at Narrabri is downstream of a distributary offtake. Changes in its efficiency through time probably account for the peculiar pattern (Fig. 2).

3. *Macquarie (Basin 421)*. Four stations can be used in the Macquarie Basin, three on the Macquarie River and one on the Bogan River (Fig. 1). They show very similar changes from 1901 to 1986 (Fig. 2). Wellington, Dubbo and Narromine are close to each other, so the pattern should be the same. The Bogan River (Peake Hill) is a tributary with its headwaters in the western slopes, and not in the eastern highlands. Cumulative deviations from the mean show flows below average from 1901 to 1914, near average to 1920, below average to 1949, a very wet period to 1954 and then near average flows ever since then.

4. *Lachlan (Basin 412)*. Along the Lachlan River, its tributaries and distributaries, the pattern of deviations from the mean for the ten stations is essentially constant (Fig. 2) and the whole of the Lachlan catchment has responded in the same way to secular climatic change. There is a dry period from 1901 to 1916, one of near average flow to the mid 1930s below average to the mid 1940s and a very wet period to the mid 1950s. There were two small wet periods in the early 1960s and mid 1970s, but flows from the mid 1960s to the present were near average.

5. *Murrumbidgee (Basin 410)*. Twenty stations were used in the Murrumbidgee catchment (Table I) and all show a minimum in the cumulative deviation from the mean in the 1940s (Fig. 2), but there are significant differences among the stations. Maximum negative values of the cumulative deviation from the mean for Wagga, Hay, Balranald, Gundagai, Narrandera and Darlington Point are much less than those for the more northerly catchments. Annual flows are less

variable downstream, possibly because the discharge of
floodwaters through distributary channels reduces the
magnitude of the peak discharge for downstream stations
without significantly affecting lower flows. Regulation
of the Murrumbidgee may also account for the decrease in
variation (see Page, this volume).

Flows are most variable at Jugiong, but because it is
not the smallest catchment (Table I), the variation cannot
be solely attributed to catchment area. The least
variable station is nearby on the Goodradigbee River at Wee
Jasper. It has a catchment area similar to that of Jugiong
Creek.

The two Upper Tumut stations, Buddong Falls and
Talbingo, show similar patterns with a marked dry between
1911 and 1918, a long period of near average flows to 1947,
followed by above average flows to 1977 and finally near
average flows. The rest, many of which are on distributaries
(offtakes), show a general pattern of below average flows
between 1901 and 1914 and then a period of minor
variations about average flows to 1947 followed by a wet
period. To varying degrees all show the 1947 minima and
also another for 1972. It appears that the early dry period
was more significant for these streams than for the
Murrumbidgee River but that the wet period between 1914 and
1923 was less significant.

6. *Murray (Basin 409)*.   The variation is less for the
Murray stations of Corowra Road bridge and Barham and
greatest at Stoney Crossing on the Wakool River (Table I,
Fig. 2), but all four stations show a similar pattern.
From 1914 to 1918, there was a short wet period followed
by near average flows to 1941. The dry period then continued to 1950 but was poorly defined compared with northern
streams. There were dry periods from 1959 to 1972 and from
1977 to the present. The Murray River is more affected by
irrigation than the other rivers.

7. *Warrego (Basin 423)*.   The two Warrego River
stations (Table I) are adjacent and patterns of the annual
flows for the main channel and distributary (Bywash) are
the same. The minimum of the deviation from the mean
occurs in 1948, similar to other stations in the Upper
Barwon basin.

8. *Darling (Basin 425)*.   Only two Darling River
stations are suitable for the analysis (Table I), these
being at Wilcannia on the main and distributary channels.

The pattern of deviations for the main channel shows a dry period between 1901 and 1947 interrupted by a small wet between 1918 and 1923. 1948 to 1954 was wet and was followed by a dry period to 1972.

B. General Trends in Annual Flows

Temporal and spatial aspects of the variation in flow can be assessed by examining the downstream trends for major sub-basins. Two indices of change are used: the magnitude of the maximum negative deviation from the mean (the larger the deviation, the greater the impact of the dry period to the 1940s) and the ratio of the mean annual flow of the two periods 1947-1985 and 1901-1946 (indicating the magnitude of the impact of the change in flow).

Whilst there is clear evidence of a dry period between the mid 1930s and late 1940s, its magnitude varied areally throughout the whole of the Murray Darling Basin. There can be no doubt that this was primarily a response to climate with flow increase in the post 1940s also attributed to a precipitation increase. However, the secular climatic change is not seen in all the flow records. Therefore other factors must be influencing its impacts.

There is a relationship between the maximum value of the negative of the cumulative deviation from the mean and catchment area for each sub-catchment (Fig. 3a). In the first instance, trend lines were defined graphically by identifying the spread of points in each basin. Second-degree polynomials were then fitted by least squares to catchments with four or more stations in a recognised cluster and with a definite minimum in the 1940s. Three regression lines are significant at the 5% level, for the Lachlan, Upper Barwon and Murrumbidgee-A. Correlation coefficients are high for the other two sub-catchments, but not significant. Effluent streams were omitted because of the difficulty in defining their catchment area. The two Murray stations fit this trend, but no trend is defined for the Macquarie River because two rivers are represented. The Murrumbidgee stations are in two distinct sets; the lower values of minima include stations along the main channel, and the second set includes stations in the upper reaches of the catchment. Negative deviations appear largest for catchment areas between 10,000 and 30,000 km^2. This suggests that the greatest change in annual flows and impact of the 1940s dry are in the middle reaches of the large basins.

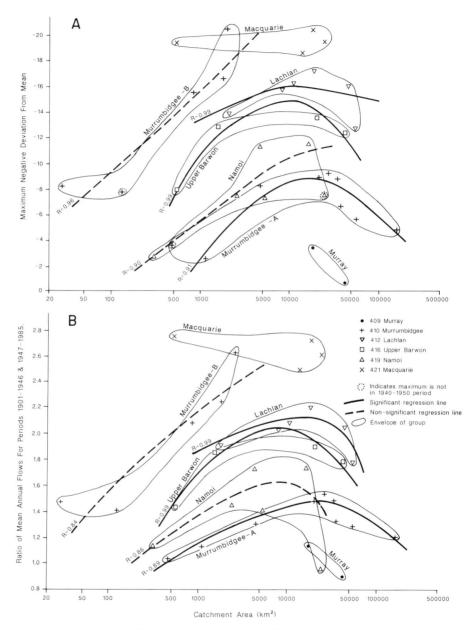

FIGURE 3. *Relationship between (a) catchment area and the maximum of the negative deviation from the mean and (b) catchment area and the ratio of the means 1946-1985 and 1901-1945.*

The Upper Barwon, Lachlan, Macquarie and Murrumbidgee-B basins have the largest minimum deviations, while the Murray has the smallest (Fig. 3a) and the Murrumbidgee-A and Namoi also have small values. Regulation of the rivers for irrigation and flood control affects annual flows, but does not explain trends observed in the northern rivers which have only recently been subject to control. Variation in the magnitude of secular climatic change over NSW is the logical explanation for the different trend lines of Figure 3a. Pittock (1983; Fig. 1) and Cornish (1977; Fig. 2) both show that the area of greatest increase in rainfall includes the Macquarie, Upper Murrumbidgee and Lachlan catchments. However, the Namoi and Upper Barwon Basins do not fit this pattern; the former has had significant secular changes in annual and seasonal rainfall, while the latter is marginal to the main area of change. The distribution of rainfall changes does not explain why the negative deviation has a maximum value for catchments between 10,000 and 30,000 km^2.

C. *Comparison of Annual Flows for 1901-1946 and 1947-1985*

For most stations the regime shift took place between 1945 and 1950 (Fig. 2). The year 1947 has been adopted to ensure uniformity in examining the change in annual flows among all the stations.

Means and standard deviations of the annual flows for the 1901-1946 are less than those for 1947-85 (Table II). The differences between the means and standard deviations for all the stations are significant at the 0.0001 level (using a t-test for non-equal variance). The ratios of mean annual flow for the 1947-1985 to 1901-1946 periods are less than 1 only at Barham (Murray) and Narrabri (Namoi) (Table II). There are no stations where the ratio of standard deviations for the two periods is less than 1.

For the large majority of stations means and standard deviations are significantly different between the two periods (using t-test for unequal variances and Cochrans test, Table II). This is not the case for the Murray stations and for several stations in the Murrumbidgee, Namoi and Warrego catchments. The variance of the annual flows is not significantly different for only five stations, one in the Murray and four in the Murrumbidgee catchments.

The relationship between catchment area and the ratio of means of the annual flows of the two periods is similar to those between area and the minima of the cumulative departures from the mean (Fig. 3b). Three of the regression lines are significant at the 5% level. The general trend

TABLE II. Means, Standard Deviations and Ratios of Means and Standard Deviations for the Periods 1901-1946 and 1947-1985.

Basin/ Station	Means of annual flows		Ratio of means	Standard deviations of flows		Ratios of stdev.	Number of years of record	
	1	2	2/1	1	2	2/1	1	2
	1901-1946	1946-1985		1901-1946	1947 1985			
	GL	GL		GL	GL		yrs	yrs
409								
002	4737.	5555.	1.17*	2069.	2813.	1.36	35	38
005	4320.	3969.	0.92*	1990.	2017.	1.01*	35	38
013	1973.	2321.	1.18*	2032.	3481.	1.71	34	38
014	967.	1037.	1.07*	647.	1032.	1.60	35	38
410								
001	3026.	4628.	1.53	1709.	2659.	1.56	46	38
002	2200.	2824.	1.28*	1457.	2376.	1.63	46	38
003	1959.	2361.	1.21*	1278.	2266.	1.77	46	38
004	2729.	4087.	1.50	1479.	2172.	1.47	46	38
005	2551.	3771.	1.48	1641.	2737.	1.67	46	38
007	248.	356.	1.44	174.	275.	1.58	46	38
009	62.	87.	1.41	35.	39.	1.11*	46	38
012	66.	86.	1.31*	80.	89.	1.11*	46	38
014	100.	137.	1.38	57.	86.	1.51	46	38
015	125.	201.	1.60	96.	197.	2.05	46	38
016	170.	242.	1.42*	157.	257.	1.64	46	38
017	181.	344.	1.90	187.	425.	2.28	46	38
018	92.	166.	1.82	97.	207.	2.14	46	38
021	2175.	2903.	1.33	1360.	2294.	1.69	46	38
024	303.	342.	1.13*	172.	214.	1.25*	46	38
025	118.	310.	2.64	159.	515.	3.23	46	38
029	16.	23.	1.49	8.	9.	1.07*	46	38
700	133.	140.	1.05*	92.	120.	1.31	46	38
701	71.	149.	2.10	70.	130.	1.85	46	38
702	96.	211.	2.21	94.	189.	2.02	46	38
412								
002	569.	1201.	2.11	603.	1168.	1.93	46	38
004	674.	1478.	2.19	698.	1502.	2.15	46	38
005	203.	360.	1.77	194.	337.	1.74	46	38
009	121.	230.	1.90	109.	220.	2.01	46	38
011	480.	1004.	2.09	533.	1047.	1.96	46	38
012	95.	204.	2.15	122.	323.	2.65	43	38
014	256.	388.	1.52	171.	262.	1.53	46	38
017	291.	481.	1.66	187.	334.	1.79	46	38
020	97.	206.	2.12	125.	326.	2.60	44	38
067	490.	1001.	2.04	463.	902.	1.95	46	38
416								
001	473.	847.	1.79	370.	750.	2.02	46	38
002	671.	1278.	1.90	551.	1032.	1.87	46	38
003	50.	73.	1.46	33.	63.	1.92	46	38
004	95.	177.	1.86	76.	133.	1.76	46	38
419								
001	566.	968.	1.71	462.	1013.	2.19	46	37
002	303.	283.	0.93*	272.	372.	1.36	46	37
004	56.	64.	1.14*	45.	49.	1.11	46	37
005	213.	307.	1.44	146.	254.	1.74	46	37
006	229.	394.	1.72	173.	386.	2.22	46	37
007	335.	468.	1.40*	256.	464.	1.82	46	37
421								
001	624.	1690.	2.71	640.	2044.	3.19	46	35
003	563.	1397.	2.48	598.	1666.	2.78	46	35
006	709.	1841.	2.60	718.	2354.	3.28	46	35
010	9.	25.	2.76	10.	36.	3.48	46	35
423								
001	7.	35.	4.78	17.	77.	4.47	25	35
002	42.	63.	1.51*	36.	70.	1.97	25	35
425								
002	1886.	4156.	2.20	1520.	5241.	3.45	46	37
008	1814.	3134.	1.73	1271.	2752.	2.16	46	38

GL Gigalitres
* differences in means and standard deviations not significant at 5% level

of high values in the Macquarie, Lachlan and Upper Darling persists. These basins also show the greatest increase in variability of flow (Table II).

VI. DISCUSSION

There can be little doubt that the secular climatic change of the 1940s caused major changes in annual flows. Those areas of NSW where the rainfall increases were also greatest where the change in annual flows was the greatest. It also seems evident that they appear to be greater in the 'middle' reaches of large basins, where rivers are passing from highlands to plains.

The Lachlan, Macquarie and Upper Barwon catchments have experienced the greatest changes. Smaller changes in the Murrumbidgee and Murray may be also influenced by regulation for flood control and irrigation. River storages should have reduced the variation in annual flows, yet all streams show some increase in variation, most being significant.

The near-parallel trend lines (Fig. 3) suggests that sub-catchments are responding in a similar manner to rainfall variation and that differences are largely related to the magnitude of the rainfall variation, which in turn governs the magnitude of the flow variation. However, the non-significance of two regressions and the variation about the lines suggests that other factors, such as groundwater loss and evaporation, cannot be ignored when assessing the causes of the variation in annual flow.

The similarity of the trend lines reflects physiographic similarities in the sub-catchments. Each has a headwater region of the eastern highlands, a western-slopes transmission zone and a distributary-dominated western plain, corresponding perhaps to the source, transfer and sink of Schumm (1977).

The spatial variation in rainfall between 1901-1946 and 1947-1985 is similar to the pattern of variation between the 1881-1901 and 1911-1940 periods. Gentilli (1971) identified a region that incorporates the Upper Barwon, Macquarie and Lachlan catchments where the decrease in rainfall between 1881-1910 and 1911-1940 was greatest. No significant change was noted in the Upper Murrumbidgee, although this could result from sampling problems. Thus, it appears that for the last 100 year period of reliable rainfall records, there are specific areas where the secular

oscillations in rainfall have had similar spatial magnitudes
on two occasions. Pittock (1975) has examined the causes
of these secular variations in terms of the Southern
Oscillation and the sub-tropical surface high-pressure belt.

Suggested causes of Holocene and Quaternary climatic
fluctuations are numerous, although the Milankovitch
hypothesis appears to explain a number of variations
(Chappell, 1978). The linkage between the secular and
longer period climatic changes is not established, but
Rognon and Williams (1977) suggest that the position and
wavelength amplitude of the subtropical western jet stream
and its influence on subtropical anticyclones is a
significant factor in the variation of precipitation (and
evaporation) in Australia. This hypothesis for longer term
climatic changes incorporates the explanation for the short
term, secular changes, namely, shifts in the wave pattern
of the general atmospheric circulation (Pittock, 1975,
p.503).

Headwater areas affected by Quaternary glaciation or
periglaciation were very small (Galloway, 1963; 1965;
Williams, 1984). Snow cover was more extensive and
some highland areas may have been subject to nivation. Thus
most of the Murray-Darling headwater region did not cross
a morpho-climatic threshold and Quaternary and Holocene
climatic changes only resulted in intensification of existing
fluvial and hillslope processes and changes in the seasons
of most intensive activity.

Williams' (1984a; 1984b) diagrams suggest that the
change zone between winter and summer rainfall for the last
25,000 years is in northerly sections of this basin. If
this is the case, it conforms with the areas of maximum
secular change in rainfall and annual flow as described.
Thus both secular and longer term climatic changes
affecting annual flows sedimentation and geomorphological
processes would probably be greatest in the northern three
sub-basins.

Changes in rainfall and annual flow may not necessarily
result in large geomorphic changes because secular
hydrologic changes are of short duration and may not last
long enough for changes to occur and for thresholds to be
exceeded in western channels (see Erskine and Warner, this
volume). However, this analysis indicates that the optimum
areas to investigate impacts may be in the Upper Barwon,
Lachlan, Upper Murrumbidgee and Macqarie catchments.

VII. CONCLUSION

Almost all the streams in the Murray-Darling basin show the impact of the 1940s secular change in rainfall. There is a distinct regional pattern in the variation of annual flows, largest in the north and in the Upper Murrumbidgee and smallest in the south, in the Murray. These changes are also largest for middle size catchments in 10,000 to 30,000 km^2 range.

The areas where secular climatic and hydrologic changes were largest between the 1901-1946 and 1947-1985 are probably the same as those where secular changes have been largest in the long term in eastern Australia. These areas are worthy of more intensive study.

ACKNOWLEDGEMENTS

The Department of Water Resources made available the data for this essay. The Central Mapping Authority is thanked for helping to obtain the data. The cartographers of the School of Earth Sciences and Margaret MacFarlane are thanked for assistance with the diagrams. Comments by colleagues, reviewers and the editor have been appreciated.

REFERENCES

Australian Academy of Science (1976) *Report of a Committee on Climatic Change*. Report No. 21.
Australian Water Resources Council (AWRC) (1984) Stream gauging information, Australia. *AWRC Water Research Series No. 2*, 6th edition.
Baker, B.W. and Wright, L.G. (1978) The Murray River: its hydrologic regime and the effects of water development on the river. *Proc. Roy. Soc. Vict.*, 90, 103-110.
Bell, F.C. and Erskine, W.D. (1981) Effects of recent increases in rainfall on floods and runoff in the upper Hunter valley. *Search*, 12, 82-83.
Brown, J.A.H. (1983) *Water 2000. Consultants Report No. 1, Australia's Surface Water Resources*, Dept. of Res. and Energy.
Chappell, J. (1978) Theories of Upper Quaternary ice ages. In A.B. Pittock, L.A. Frakes, D. Jenssen, J.A. Peterson and J.W. Zillman (eds.) *Climatic Change and Variability*, Cambridge Univ. Press, 211-224.

Collett, K.O. (1978)   The present salinity position in the
    River Murray Basin. *Proc. Roy. Soc. Vict.,,* 90, 111-
    124.
Cole, P.J. and McCloud, P.I. (1981)   The effect of River
    Murray salinity on citrus production. *AWRC Tech. Pap.,*
    No. 62.
Cornish, P.M. (1977)   Changes in seasonal and annual rainfall
    in New South Wales. *Search*, 8, 38-40.
Crabb, P. and Riley, S.J. (1987)   Surface water resources.
    In R.J. Harriman and E.S. Clifford (eds.) *Atlas of
    NSW*, Central Mapping Auth. NSW.
Department of National Development (1978)   Variability of
    runoff in Australia. *AWRC Hydrological Series*, No. 11.
Erskine, W.D. (1986)   River metamorphosis and environmental
    change in the MacDonald Valley, New South Wales, since
    1949. *Aust. Geog. Studies*, 24, 88-107.
Erskine, W.D. and Bell, F.C. (1982)   Rainfall, floods and
    river channel changes in the upper Hunter. *Aust. Geog.
    Studies*, 20, 183-196.
Galloway, R.W. (1963)   Glaciation in the Snowy Mountains:
    a reappraisal. *Proc. Linn. Soc. NSW,* 88, 180-198.
Galloway, R.W. (1965)   Late Quaternary climates in
    Australia. *J. Geol.,* 73, 603-618.
Gentilli, J. (1971)   Climatic   fluctuations.   In J.
    Gentilli (ed.) *Climates of Australia and New Zealand.
    World Survey of Climatology*, Vol. 13, Elsevier,
    Amsterdam, 189-211.
Goldsmith, E. and Hilgard, N. (1984) *The Social and
    Environmental Effects of Large Dams*, Vol. 1 Overview.
    Wadebridge Ecol. Centre, Cornwell.
Henry, H.M. (1977) Catastrophic channel changes in the
    MacDonald Valley, NSW, 1949-1955. *J. Roy. Soc. NSW*,
    110, 1-16.
Hills, E.S. (1972) *Elements of Structural Geology*, Chapman
    and Hall, Melbourne.
Kraus, E.B. (1954)   Secular changes in the rainfall regime
    of SE Australia. *Quart. J. Roy. Met. Soc.*, 80, 591-601.
Lettenmaier, D.P. and Burgess, S.J. (1978)   Climatic
    change - detection and its impact on hydrological
    design. *Water Resources Res.*, 14, 679-687.
Nanson, G.C. (1986) Episodes of vertical accretion and
    catastrophic stripping: a model of disequilibrium
    flood-plain development. *Bull. Geol. Soc. Am.*,
    97, 1467-1475.

Neller, R.J. (1980) Channel change in the Macquarie
    Rivulet, New South Wales. *Zeit. f. Geomorph.*, 24(2),
    168-179.
New South Wales Dams Safety Committee (1985) *Annual
    Report 1983-84.*
Ollier, C.D. (1978) Tectonics and geomorphology of the
    eastern highlands. In J.L. Davies and M.A.J. Williams
    (eds.) *Landform Evolution in Australasia*, ANU Press,
    Canberra, 5-47.
Petts, G.E. and Lewin, J. (1979) Physical effects of
    reservoirs on river systems. In G.E. Hollis (ed.)
    *Man's Impact on the Hydrological Cycle in the United
    Kingdom*, Geobooks, Norwich, 79-91.
Pickup, G. (1976) Geomorphic effects of changes in river
    runoff, Cumerland Basin, NSW. *Aust. Geog.*, 13, 188-193.
Pickup, G. and Warner, R.F. (1976) Effects of hydrologic
    regime on magnitude and frequency of dominant
    discharges. *J. Hydrol.*, 29, 51-75.
Pigram, J.J. (1986) *Issues in the Management of Australia's
    Water Resources*, Longman, Melbourne.
Pittock, A.B. (1975) Climatic change and the patterns of
    variation in Australian rainfall. *Search*, 6, 498-504.
Pittock, A.D. (1978) Pattern of variability in relation
    to the general circulation. In A.B. Pittock, L.A.
    Frakes, D. Jenssen, J.A. Peterson and J.W. Zillman
    (eds.) *Climatic Change and Variability: A Southern
    Perspective*, Cambridge Univ. Press, 167-179.
Pittock, A.B. (1983) Recent climatic change in Australia:
    implications for a CO_2 warmed earth. *Clim. Change*,
    5, 321-340.
Riley, S.J. (1981) The relative influence of dams and
    secular climatic change on downstream flooding,
    Australia. *Water Res. Bull.*, 17, 361-366.
River Murray Commission (1970) *Murray Valley Salinity
    Investigation*, Gutteridge Haskins and Davey, with
    Hunting Technical Services, 2 vols.
Rognon, P. and Williams, M.A.J. (1977) Late Quaternary
    climatic changes in Australia and North Africa: a
    preliminary interpretation. *Palaeogeog., Palaeoclimat.,
    Palaeoecol.*, 21, 285-327.
Schumm, S.A. (1977) *The Fluvial System*, Wiley, New York.
Water Conservation and Irrigation Commission (WC & IC)
    (1968-1975) *Survey of Thirty-Two NSW River Valleys*
    (revelant basins).
Water Conservation and Irrigation Commission (WC & IC)
    (1971) *Water Resources of NSW*, 2 vols.

Water Resources Commission (WRC) (1982-1984) *New South Wales Inland Rivers Floodplain Management Studies*, (relevant basins).

Williams, M.A.J. (1984a) Quaternary environments. In J.J. Veevers (ed.) *Phanerozoic Earth History of Australia*, Clarendon Press, Oxford, 42-47.

Williams, M.A.J. (1984b) Cenozoic evolution of arid Australia. In H.G. Cogger and E.E. Cameron (eds.) *Arid Australia*, Australian Museum, Sydney, 59-78.

Woodyer, K.D. (1978) Sediment regime of the Darling River. *Proc. Roy. Soc. Vict.*, 90, 139-148.

13
Bankfull Discharge Frequency for the Murrumbidgee River, New South Wales

K.J. Page

School of Applied Science
Riverina-Murray Institute of Higher Education
Wagga Wagga

I. INTRODUCTION

The adjustment of channel morphology to the volume of water the channel must carry is a basic axiom of modern fluvial geomorphology. Following the pioneer study of Leopold and Maddock (1953) which demonstrated that channel width and depth increase downstream as power functions of increasing discharge, many studies have verified the general nature of these relationships. In the 1950s, Wolman and Leopold (1957) emphasised the special significance of channel-filling, or bankfull, discharge. They found that for streams with well-defined floodplains bankfull stage appeared to accommodate flows of between one and two years return period on the annual flood series in geographically diverse areas. Dury (1968) in turn proposed that, as a working hypothesis, bankfull discharge be assigned a recurrence interval of 1.58 years (the most probable annual flood) on the annual series. The corollary of this assumption was that stream discharge could be estimated with a high degree of confidence from the characteristics of channels ranging in size from laboratory flumes to large rivers (Dury, 1976).

Although many studies in the 1950s and 1960s found support for the constant frequency of bankfull discharge (Nixon, 1959; Dury, 1965; Kilpatrick and Barnes, 1964; Woodyer, 1968), the trend in more recent years has been to an increasing awareness of departures from the one to two

year rule (Harvey, 1969; Ackers and Charlton, 1970; Lewin
and Manton, 1975; Pickup and Warner, 1976; Williams, 1978).
In the United States, Williams (1978) examined 36 active
floodplains and found bankfull recurrence intervals
ranging from 1.01 to 32 years on the annual series. Although
the modal recurrence interval was 1.5 years, only 62 percent
of the recurrence intervals lay between one and two years.
Because of the wide range in recurrence intervals and the
considerable spread of the distribution Williams (1978)
concluded that an average recurrence interval has little
meaning for the active floodplains studied and is a poor
estimate of bankfull discharge.

    It is perhaps surprising that a study of North American
rivers should produce a finding so diametrically opposed to
Wolman and Leopold's (1957) original thesis. An
examination of both data sets suggests, however, that the
difference is essentially one of interpretation. Table I
summarises recurrence intervals of bankfull discharge
estimated by Wolman and Lepold (1957), Leopold *et al.* (1964)
and Williams (1978). Differences between the data sets are
not great. If anything, William's data show a slightly
higher proportion of rivers in the 1.2 to 2.0 year category.
Perhaps the most acceptable interpretation of Wolman and
Leopold's (1957) claim for a constant frequency of bankfull
discharge is that on meandering rivers with well-developed
floodplains, bankfull discharge frequencies are commonly
found to lie between one and two years on the annual series,
but that individual rivers may, for a variety of reasons,
possess bankfull frequencies outside this range.

*TABLE I.   Frequency of Bankfull Discharge - Selected
North American Data*

Study	Number of Observations	Annual Series Return Periods (Years)		
		<1.2	*1.2-2.0*	*>2.0*
Wolman and Leopold (1957)	*37*	*41%*	*24%*	*35%*
Leopold, Wolman and Miller (1964)	*20*	*45%*	*35%*	*20%*
Williams (1978)	*36*	*23%*	*39%*	*38%*

In Australia, Woodyer's (1968) survey of rivers in New South Wales remains the most comprehensive summary of local bankfull discharge frequencies. Woodyer considered that many streams had recently incised their channels creating terraces out of recent floodplains. In these instances he correlated bankfull stage with depositional benches located on the channel banks. Given a judicious selection of actual floodplains and benches, Woodyer found that a single frequency distribution adequately described the recurrence intervals of bankfull discharge on rivers in New South Wales. When a 16 year common period was used, all recurrence intervals fell within the range 1.24 to 1.62 years on the annual series and 0.60 and 1.04 years on the partial duration series.

Subsequent studies in eastern Australia have failed to support either Woodyer's method of defining bankfull discharge or his remarkably consistent results. An extensive survey of 47 small streams in the Cumberland Basin by Pickup and Warner (1976) found that the recurrence interval of bankfull discharge lay between four and seven years on the annual series. In studies of the New South Wales coastal region Nanson and Young (1981) and Nanson (1986) found wide discrepancies between bankfull discharge frequencies along individual streams. Their research also showed, however, that these streams are produced predominantly by vertical rather than lateral accretion, and therefore tend to progressively increase in elevation. The combination of vertical accretion and a high episodic and spatially disjunct pattern of floodplain formation was thought to account for the high variability of bankfull recurrence intervals.

These coastal rivers with floodplains dominated by vertical accretion bear little relationship to those of the westward-flowing Murray system of inland south-eastern Australia where well-defined meandering channels and floodplains dominated by scroll bar and concave bench arrays (Bowler, 1978; Page and Nanson, 1982) are typical. These floodplains conform closely to the classical lateral accretion models of floodplain formation proposed by Wolman and Leopold (1957) and Allen (1965) and permit the relatively straight forward definition of bankfull stage in the field in terms of mean floodplain elevation.

The aim of the present study was to establish bankfull discharge frequencies along an 80 km reach of the Murrumbidgee River near Wagga Wagga, New South Wales (Fig. 1). In particular it sought to determine whether a fixed frequency of bankfull discharge occurred along the study

reach and, if so, whether that frequency fell within the
limits established by Wolman and Leopold (1957) and Woodyer
(1968).

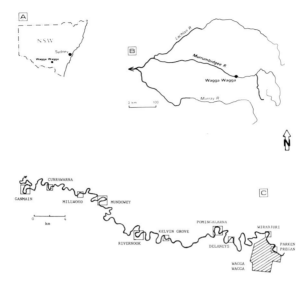

*FIGURE 1. The Murrumbidgee River study reach near Wagga
Wagga, New South Wales, Australia.*

II.   THE MURRUMBIDGEE RIVER

The Murrumbidgee River rises in the South-Eastern
Highlands of New South Wales and has a catchment area of
27700 km^2 at Wagga Wagga.   It flows generally westwards
between the Lachlan Valley to the north and the Murray Valley
to the south and has a mean annual discharge of
approximately 120 m^{3}s^{-1} at Wagga Wagga.   In common with
many of the west-flowing rivers of New South Wales, the
Murrumbidgee's flow regime has been affected by
engineering works in its upper catchment region.   The
largest water storages are Burrinjuck Dam (1,032,000 ML)
and Blowering Dam (1,632,000 ML).   The combined effect of
these and other works has been to increase summer flows for
irrigation and to attenuate the magnitude of small to
moderate floods (Page, 1979).
     The 80 km study reach extends from Wagga Wagga down-
stream to Ganmain Station (Fig. 1). Within this reach the

Murrumbidgee has a well-defined suspended-load channel with steep banks. Bankfull channel width averages 80 m, depth 6 m, slope 0.0002 and sinuosity 2.1. Lateral migration of the channel has produced a characteristic scroll and concave bench patterned floodplain (Nanson and Page, 1983). A preliminary survey indicated that the scrolled floodplain surface and the somewhat lower concave bench surface were inundated with frequencies of between 1.5 and 2.0 years, and between 1.0 and 1.5 years, respectively. Although point bars and cut banks are prominent features of the reach, the rate of lateral channel migration under the present flow regime is low. Sequential air photographs since 1941 indicate a maximum rate of cut bank retreat of around 0.4 my^{-1}, but most bends exhibit very much lower rates than this.

Near Wagga Wagga scrolled floodplain deposits form a 6-7 m layer of sands and muds overlying coarse sands and gravels. At Pomingalarna (Fig. 1) bore holes commissioned by Pioneer Concrete (M. Quirk, pers. comm.) showed that the basal gravels extend to at least 15 m below the present river bed. The gravels armour the river bed and are essentially immobile under the present flow regime. Above the gravels the alluvium changes vertically from point bar sands to cohesive fine sandy silts and clays deposited from overbank flows. Meander cutoffs with varying thicknesses of silty clay and clay infill are common features and play an important role in the attenuation of flood peaks along the study reach.

III. FIELD DEFINITION OF BANKFULL STAGE

By definition bankfull stage is reached when water just begins to overflow the banks of the channel. On the Murrumbidgee River this level varies greatly from one location to another, depending upon the configuration of the floodplain ridge and swale array. Aerial observations at rising flood stage have shown that on many bends flood water enters· swales, typically at the downstream region of each bend, before overtopping the ridges adjacent to the channel at the bend axis (Fig. 2). Under these conditions equating of bankfull stage with mean floodplain height appeared to provide the most appropriate method of averaging local variations in bank height. Although many different definitions of bankfull stage have been developed for rivers with poorly developed floodplains (Wolman, 1955; Nixon, 1959; Schumm, 1960; Woodyer, 1968; Harvey, 1969; Riley,

1972), there appears to be little basis for departing from Wolman and Leopold's (1957) original definition in terms of mean floodplain elevation where a well-defined surface of lateral accretion exists.   In this study bankfull stage at a given cross section was defined by the mean elevation of the first three scroll ridges and intervening swales adjacent to the present channel.

Altogether ten sites with well-developed floodplains of lateral accretion were surveyed at more or less regular intervals along the study reach (Fig. 1).   Cross sections were surveyed at right angles to the local ridge and swale array on the convex (point bar) side of each bend.   At least two sections were surveyed in the region of the bend axis at each site.   On larger, more complex bends (for example, Pomingalarna and Ganmain) additional sections were surveyed to provide a more reliable estimate of mean ridge and swale elevation.   All sections were surveyed at times of low river flow (less than 2.0 m at Wagga Wagga gauge) with vertical readings to 0.01 m.   By noting river height on the day of survey all levels could subsequently be related to Wagga Wagga gauge datum by assuming a constant river slope between Wagga Wagga and Mundowey and between Mundowey and Narrandera.

FIGURE 2.   Oblique air photographs of Rivernook and Mundowey bends showing the scroll-patterned floodplain and floodwater entering swales at the downstream region of each bend.

IV.   FLOOD FREQUENCIES ON THE STUDY REACH

The discharge frequency of bankfull stage on a river is usually determined by means of a field survey of the channel and adjacent floodplain at a site nearby a suitable river

gauge. A given river stage can then readily be converted
to a flood frequency by reference to the appropriate rating
curve. For sites between gauges, however, it is necessary
to estimate flood frequencies by interpolation.

On the Murrumbidgee River flood peaks decline markedly
downstream of Wagga Wagga. Page and McElroy (1981) showed
on the basis of a 40 year common period of record that the
mean annual flood ($Q_{2.33}$) at Wagga Wagga is 59,000 ML day^{-1}
compared to only 37,500 ML day^{-1} at Narrandera. For larger
floods the reduction is even greater amounting to more than
40 percent at the five year flood. This downstream decline
in flood peaks is produced largely by a combination of high
floodplain storage and minimal tributary input.

To estimate bankfull frequencies downstream of Wagga
Wagga, it was therefore necessary to estimate the rate of
flood peak decline along the study reach. Initially flood
magnitude-frequency data were calculated from gauge records
at Wagga Wagga and Narrandera for the period 1937 to 1976
inclusive (Page and McElroy, 1981). Discharges so
calculated were then converted to river stages by reference
to Department of Water Resources rating tables. A flood-
warning station at Mundowey Station (Fig. 1) provided a
partial record of flood peaks between 1950 and 1984 and
hence, a basis for the comparison of flood stages at Wagga
Wagga and the downstream part of the study reach.

Figure 3 shows peak stages at Wagga Wagga and Mundowey
of 26 floods between 1950 and 1984 (data provided by Mr C.
Cameron, Mundowey Station). As expected the data are highly
correlated ($r = 0.97$, $p < .001$) and show minimal scatter. A
trend line fitted by the method of principal axes (Sokal
and Rohlf, 1981) permitted the conversion of flood stage at
Wagga Wagga to flood stage at Mundowey. To adjust for
differences in gauge datum between the two sites, 0.7 m must
be subtracted from all Mundowey figures. This represents
the difference between low-flow readings at the two gauges
during periods of nearly constant flow. For example, a
gauge reading of 1.4 m applies to a flow of 5180 ML day^{-1} at
Wagga Wagga. At Mundowey this flow has a stage of 2.1 m.
A similar adjustment to readings at Narrandera requires a
deduction of 1.4 m. Figure 4 shows the pattern of
declining stage between Wagga Wagga and Narrandera for
selected annual floods.

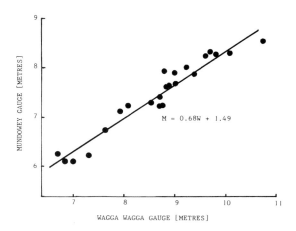

FIGURE 3.   *Peak stages of 26 floods at Wagga Wagga and Mundowey Gauges between 1950 and 1984.*

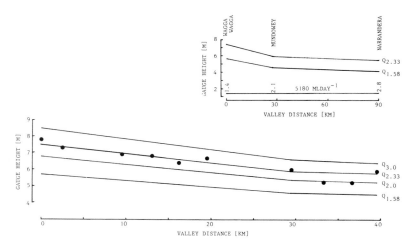

FIGURE 4.   *(Upper) Declining stages of $Q_{1.58}$ and $Q_{2.33}$ between Wagga Wagga and Narrandera. Base line shows the stage of 5180 ML day^{-1} discharge at Wagga Wagga, Mundowey and Narrandera. (Lower) Relationship between bankfull stage (solid dots), selected flood frequencies and distance down valley along the study reach.*

V.   RESULTS

Representative cross sections at each of the ten field sites are shown in Figure 5. Although considerable between section variation in scroll spacing and local relief exists, the strongly corrugated nature of the floodplain surface is apparent at all sites. In general the sections show little progressive change in mean elevation away from the present channel. Table II summarises data for bankfull stage, bankfull recurrence interval and mean annual flood stage at each site. Mean values of bankfull stage were then plotted against valley distance downstream along the reach (Fig. 4). Also plotted were the stages of selected annual floods with return periods ranging from 1.58 to 3.0 years on the annual series. The partial duration series return periods which correspond to actual time periods were established from data in Page and McElroy (1981).

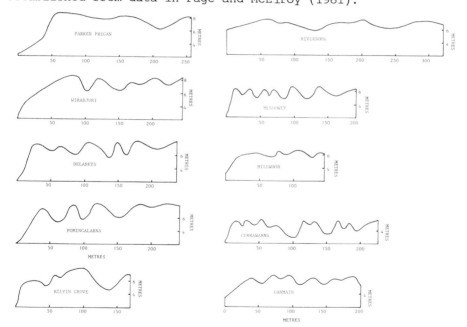

FIGURE 5. Representative cross sections of the scroll-patterned floodplain at each field site.

Annual series recurrence intervals of bankfull stage along the study reach range from 1.95 to 2.45 years. The mean value of 2.27 years and standard deviation of 0.18 years

TABLE II. Stage and Frequency of Bankfull and Mean Annual Flood Discharge on Wagga Wagga reach of Murrumbidgee River.

Site	Bankfull Stage (m)	Recurrence Interval Annual Series (y)	Partial Series (y)	$Q_{2.33}$ Stage (m)
Parken Pregan	7.80	2.45	1.55	7.50
Wiradjuri	7.30	2.30	1.40	7.40
Delaneys	6.90	2.30	1.40	7.00
Pomingalarna	6.80	2.30	1.40	6.90
Kelvin Grove	6.40	2.20	1.30	6.60
Rivernook	6.70	2.45	1.55	6.45
Mundowey	6.00	2.40	1.50	5.90
Millwood	5.25	1.95	1.05	5.85
Currawarna	5.20	1.95	1.05	5.85
Ganmain	5.90	2.35	1.45	5.80
Mean		2.27	1.37	
Standard Deviation		0.18	0.18	

show that the data conform closely to the mean annual flood which has a recurrence interval of 2.33 years. The partial duration series mean of 1.37 years shows that on average bankfull stage can be expected to be equalled or exceeded more than twice every three years. The close relationship between bankfull stage and declining stage of mean annual flood along the reach is supported statistically by a product moment correlation coefficient of 0.94 ($p < 0.001$) and minimal data scatter. The mean square of deviations of bankfull stage from mean annual flood stage is only 0.1. These data suggest that the frequency of bankfull stage is almost constant along the study reach and that floodplain elevation is adjusted to the contemporary flood regime.

VI. DISCUSSION

Although the present results support a nearly constant frequency of bankfull discharge along the study reach, the actual frequency is rather less than is considered typical by Wolman and Leopold (1957), Dury (1968) or Woodyer (1968). Of course it is now clear that the frequency of bankfull

discharge is neither literally nor statistically constant
in diverse geographic regions (Williams, 1978). Floodplains
dominated by vertical accretion in particular exhibit
highly variable bankfull frequencies even along short reaches
of a single channel (Nanson, 1986). Where meandering
channels and floodplains of lateral accretion occur, a
frequency of one to two years on the annual series is common
but not universal. Indeed, an examination of Wolman and
Leopold's (1957) data suggests that they may have overstated
the case for the constancy of bankfull stage for their
results differ little from those of Williams (1978), who
arrived at an almost diametrically opposed conclusion. Of
all the studies of bankfull discharge frequency only Woodyer
(1968) supported the claim for a single frequency
distribution with a rigorous statistical analysis. In
general, detailed studies of bankfull discharge along
individual rivers are relatively uncommon, probably because
of the difficulty of estimating flood frequencies accurately
at sites distant from river gauges.

Because the flow regime of the Murrumbidgee River has
been altered since 1907 by reservoir construction and later
by flow augmentation, it remains to be seen whether the
findings of the present study are truly at odds with the 1
to 2 year rule for bankfull discharge frequency. For
instance, Burrinjuck Dam was constructed upstream in stages
between 1907 and 1956. Irrigation water was first
supplied in 1912. Major reservoir construction also
occurred on the Tumut River between 1956 and 1971. It is
generally recognised that the construction of such large
storages tends to reduce the size of flood peaks along
downstream reaches. In considering the effects of
Burrinjuck Dam upon lower Murrimbudgee Valley flooding,
Sinclair Knight and Partners (Water Resources Commission,
1977) noted a significant mitigating effect for moderate
floods in which flow volume is comparable to dam volume,
but a negligible effect for major floods. This volume of
Burrinjuck Dam (1,030,000 ML) is approximately equivalent
to the water flux of a six to ten year flood event with a
peak flow of between 120,000 and 150,000 ML day^{-1} at the
dam. Clearly then, floods with recurrence intervals of less
than around five years are likely to have been modified by
dam construction.

Flow records have been kept at Wagga Wagga since 1885.
They provide a useful basis for the examination of the
Murrumbidgee flow regime before reservoir construction
commenced in 1907. Figure 6 shows magnitude-frequency plots
of annual floods on the Murrumbidgee at Wagga Wagga from

1885 to 1906 and 1937 to 1976. The data indicate a mean annual flood of 79,000 ML day^{-1} and a most probable annual flood of 57,000 ML day^{-1} in the early period compared to discharges of 58,000 ML day^{-1} and 37,500 ML day^{-1} respectively for the period 1937 to 1976. Both these periods incorporate flows from both flood- and drought-dominated regimes but are mainly flood dominated (eg, 1885-1900 and 1946-1976) (see Erskine and Warner, this volume).

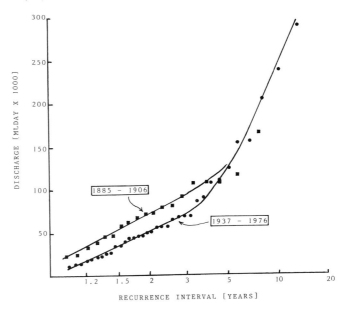

FIGURE 6. Magnitude and frequency of annual floods on the Murrumbidgee River at Wagga Wagga from 1885 to 1906 and from 1937 and 1976.

The evidence of maps, sequential air photographs since 1944 and shifts in rating at river gauges all indicate that the Murrumbidgee River has been essentially stable in the past century and therefore quite unlike some of the coastal rivers (Erskine and Warner, this volume; Pickup, 1976). Near Wagga Wagga the channel is characterised by low rates of lateral migration at bends with rates of cut bank erosion only rarely exceeding 0.2 my^{-1} and minimal scroll bar accretion. Typically, the most recent scrolls support mature river red gums conservatively estimated to be at least 100 years old.

The inescapable conclusion is therefore that the

frequency of flooding on the Murrumbidgee has been reduced mainly be reservoir construction and that the channel has adjusted little to the changed flow regime. In other words the Murrumbidgee floodplain appears to be adjusted to a flood regime in which bankfull stage was attained rather more frequently than it is at present. On the basis of current rating tables, it seems likely that the present mean annual flood had a recurrence interval of little more than 1.6 years before 1907. This change is equivalent to a reduction in the stage of the mean annual flood by approximately 1 m at Wagga Wagga and places the original frequency of bankfull stage squarely in the one to two year interval on the annual series. Of course, a return period of 1.6 years on the annual series corresponds to an actual return period of less than one year (Page and McElroy, 1981).

The finding of this survey of the Murrumbidgee River near Wagga Wagga has clear implications for studies of flood frequencies of other rivers, both in the Murray system and elsewhere, where flow regimes have been significantly altered by regulatory structures in comparatively recent times. Low contemporary frequencies of bankfull stage in such cases may well reflect engineering impacts on flood regimes, rather than any fundamental catchment relationship between channel capacity and floodplain elevation (Riley, this volume).

REFERENCES

Ackers, P. and Charlton, F.G. (1970) Meander geometry arising from varying flows. *J. Hydrol.*, 11, 230-252.
Allen, J.R.L. (1965) A review of the origin and character of recent alluvial sediments. *Sedimentol.*, 5, 89-191.
Bowler, J.M. (1978) Quaternary climatic change and tectonics in the evolution of the Riverine Plain, South Eastern Australia. In J.L. Davies and M.A.J. Williams, (eds.). *Landform Evolution in Australia*, A.N.U. Press, Canberra.
Dury, G.H. (1965) Theoretical implications of underfit streams. *U.S.G.S. Prof. Paper, 452-C.*
Dury, G.H. (1968) Bankfull discharge and the magnitude frequency series. *Aust. J. Sci.*, 26, 371.
Dury, G.H. (1976) Discharge prediction, present and former, from channel dimensions. *J. Hydrol.*, 30, 219-245.

Harvey, A.M. (1969) Channel capacity and the adjustment of streams to hydrological regime. *J. Hydrol.*, 8, 82-98.

Kilpatrick, F.A. and Barnes, H.H. (1964) Channel geometry of piedmont streams as related to frequency of floods. *U.S.G.S. Prof. Paper 422-F.*

Leopold, L.B. and Maddock, T. (1953) The hydraulic geometry of stream channels and some physiographic implications *U.S.G.S. Prof. Paper 252.*

Leopold, L.B., Wolman, M.G. and Miller, J.P. (1964) *Fluvial Processes in Geomorphology*, Freeman, San Francisco.

Lewin, J. and Manton, M.M.M. (1975) Welsh floodplain studies: the nature of floodplain geometry. *J. Hydrol.*, 25, 37-50.

Nanson, G.C. (1986) Episodes of vertical accretion and catastrophic stripping: a model of disequilibrium floodplain development. *Geol. Soc. Amer. Bull.*, 97, 1467-1475.

Nanson, G.C. and Page, K.J. (1983) Lateral accretion of fine grained concave benches on meandering rivers. *Spec. Publs. Int. Ass. Sedimentol.*, 6, 133-143.

Nanson, G.C. and Young, R.W. (1981) Overbank deposition and floodplain formation on small coastal streams of New South Wales. *Zeit. f. Geomorph.*, 24, 332-347.

Nixon, M. (1959) A study of the bankfull discharges of rivers in England and Wales. *Proc. Inst. Civil Engrs.*, 12, 157-175.

Page, K.J. (1979) Altered hydrologic regime on the Tumut River, New South Wales. *Geog. Bull.*, 11, 133-139.

Page, K.J. and McElroy, L. (1981) Comparison of annual and partial duration series floods on the Murrumbidgee River. *Water Res. Bull.*, 17, 286-289.

Page, K.J. and Nanson, G.C. (1982) Concave-bank benches and associated floodplain formation. *Earth Surf. Proc. and Landforms*, 7, 529-543.

Pickup, G. (1976) Geomorphic effects of changes in river runoff, Cumberland Basin, N.S.W. *Aust. Geog.*, 13, 188-193.

Pickup, G. and Warner, R.F. (1976) Effects of hydrologic regime on magnitude and frequency of dominant discharge. *J. Hydrol.*, 29, 51-75.

Riley, S.R. (1972) A comparison of morphometric measures of bankfull. *J. Hydrol.*, 17, 23-31.

Schumm, S.A. (1960) The shape of alluvial channels in relation to sediment type. *U.S.G.S. Prof. Paper*, 352-B.

Sokal, R.R. and Rohlf, F.J. (1981) *Biometry*, Freeman, San Francisco.

Water Resources Commission (1977) *Murrumbidgee River Flood Mitigation Study*, Sinclair Knight and Partners.

Williams, G.P. (1978) Bank-full discharge of rivers. *Water Resour. Res.*, 14, 1141-1154.

Wolman, M.G. (1955) The natural channel of Brandywine Creek, Pennsylvania. *U.S.G.S. Prof. Paper 271.*

Wolman, M.G. and Leopold, L.B. (1957) River flood plains: some observations on their formation. *U.S.G.S. Prof. Paper 282-C.*

Woodyer, K.D. (1968) Bankfull frequency in rivers. *J. Hydrol.*, 6, 114-142.

14

Problems in Assessing the Impact of Different Forestry Practices on Coastal Catchments in New South Wales

L.J. Olive
W.A. Rieger

Department of Geography and Oceanography
University College
University of New South Wales
Australian Defence Force Academy
Campbell,   ACT

I.   INTRODUCTION

In recent years, considerable research interest has been centred in Australia on investigating the influence of various forestry practices and fire on hydrology, water quality and erosion and sediment transport in catchments vegetated by native eucalypt forests. In many cases such projects have represented the most intensive and longest studies of Australian fluvial systems. Much of this work has been prompted by increasing public debate and concern on the impact of forestry operations on native forests and an awareness by foresters of the need to reduce any long-term degradation of forested catchments. The influence of wildfire, relatively common in many forested areas, has also been studied.

Many of these areas represent important water-supply catchments and any changes to water yields or quality can be of concern. Undisturbed native forested catchments generally have low suspended sediment concentrations, even by Australian standards (Olive and Walker, 1982). Any increases due to forestry operations or wildfire can have an important influence on water quality, and on land degradation through erosion. The results from these studies may also have wider application in understanding the influence of other non-point disturbances in catchments. This essay reviews the results of research in Australian forested catchments and, using results of work carried out by the authors on sediment transport and erosion in the Eden hydrology project in southern New South Wales, examines the wider implications of the research.

II.   STUDIES OF FORESTED CATCHMENTS IN AUSTRALIA

Most of the research in Australia has adopted the paired catchment approach developed in North America where many studies have been carried out examining the influence of forestry operations (see articles in Davies and Pearce, 1981 and O'Loughlin and Pearce, 1984). In this approach a number of experimental catchments are monitored and, after a period of calibration, various forestry operations such as roading and logging are carried out on some of the catchments while at least one is left in an undisturbed state to act as a control. Comparisons are then made of the hydrology, water quality and sediment concentrations and loads before and after treatment in the disturbed catchments using the control catchment(s) as a baseline. This represented a major change in approach in Australia, because previous research, particularly with respect to sediment transport, had been based on short-term thesis-type studies (Douglas, 1973; Loughran, 1984). The forestry projects are larger, last longer and are more detailed studies.

A.   *Studies of Forestry Operations*

Research projects have been established in most states to examine the impact of various forestry operations on streams, often initiated by state Forestry Commissions or water authorities. Many have used an integrated approach with several groups of researchers involved in particular projects. Much of the work has concentrated on water yields and hydrologic response but increasing emphasis is being placed on water quality and erosion. This review concentrates on published material but reference is made to other more recent projects where results are not yet published.

The best known and longest-running project in the country is the Melbourne and Metropolitan Board of Works Coranderrk and North Maroondah Experiments in Victoria. These were established using the paired catchment approach in the late 1960s to examine the impact of a range of timber harvesting techniques on the yield and quality of water in mountain ash (*E. regnans*) forests which are common in many of Melbourne's water supply catchments. The catchment characteristics, experimental design and some early results are outlined in Langford and O'Shaughnessy (1979) and (1980). Langford *et al* (1982) reported on the effects of roading and harvesting of a mature mountain ash forest on water yield and quality. Peak storm discharges and water yield increased following treatment particularly in the clearfall areas; this was attributed to reduced canopy interception. After six years catchments had returned to near pre-treatment levels. Little change was observed in water chemistry but suspended sediment concentrations increased, from an average of 14 mg L^{-1} pre-treatment to 20 mg L^{-1} following clearfelling, and the increase persisted for at least six years. Suspended sediment loads increased, reflecting the increases in both discharge and sediment concentration with the major source of this being attributed to road construction.

Another project in Victoria has been centred on small catchments in the East Kiewa River, where alpine ash (*E.delegatensis*) forest has been logged (Leitch, 1982). Other isolated research projects have been carried out (Bren and Leitch, 1985, 1986).

In Queensland several workers have studied logging operations in forests including rainforest in higher rainfall areas. Much of this work is reported in detail by Bonell (this volume). Gilmour (1971) reported large increases in suspended sediment concentration following logging of a virgin rainforest catchment with sediment source areas restricted to disturbed locations particularly snig tracks and undrained roads.

In the Babinda area of north Queensland the impact of rainforest logging on streams has been reported by Gilmour (1977) and Gilmour *et al* (1982). They found statistically significant increases in water yield and there was up to a three-fold increase in suspended sediment concentrations following logging, while complete clearing resulted in larger increases. Up to eight years after logging sediment concentrations had still not returned to pre-disturbance levels. The sediment levels were closely related to vegetative cover and the condition of the soil surface. Later work has concentrated on process studies to explain water pathways and runoff generation, and to examine the role of these in determining erodibility (Bonell, this volume).

The problems of water yield and salinity have been the

major concern in Western Australia, with Stokes and Loh (1982) outlining changes in stream chloride levels following deforestation and explaining them in terms of runoff generation. Increases in both water yield and salt in the deforested catchment were the result of increased discharge magnitude and frequency rather than any major change in process. In the same catchments Sharma *et al* (1982) have considered soil water and ground water responses.

The Eden hydrology project initiated by the N.S.W. Forestry Commission in 1977 and involving several other researchers represents the major research initiative in New South Wales with a paired catchment experiment to examine the impact on small streams of clearfall logging operations in dry sclerophyll forest. Early work concentrated on quantifying changes in storm discharge response (Mackay and Cornish, 1982) and suspended sediment concentrations and loads (Rieger *et al*, 1979; Burgess *et al*, 1981). In all cases increases were recognised but were difficult to quantify. Cornish and Binns (1988) found little variation in turbidity following logging, while Mackay and Robinson (1987) reported little change in streamwater chemistry. Later work has tended to concentrate on consideration of processes involved in runoff generation (Moore *et al*, 1986) and sediment transport (Olive and Rieger, 1985, 1987; Rieger and Olive, 1986) and this work is continuing.

A further experiment is in progress in the Karuah River where eight catchments have been studied, again using the paired catchment approach and a range of logging types (Cornish, 1980). Little work from this project has yet been published. However, Riley (1984) found changes in soil permeability following logging and roading.

This review is by no means exhaustive but does outline the major research projects. In virtually all of these, there has been a common trend where initially research was based on traditional techniques developed in the USA and was aimed at quantifying changes induced by logging and its associated disturbances. With a realisation of the limitations of these techniques, particularly with respect to Australian conditions, later research has become much more process oriented in an attempt to understand the fluvial system and how any disturbances change this system.

B. Studies of Wildfire

Wildfire is a relatively common phenomenon in Australian forests and can cause major disturbances to vegetation and ground cover. Some attention has been paid to the impact

of fire on stream hydrology and erosion together with the
persistence of this influence through the regeneration phase.
Unlike studies of logging operations, these have generally
been initiated in the post-fire period which leads to some
difficulties in interpreting results, as there is commonly
no pre-fire data in burnt catchments. Such research has
been carried out by Brown (1972) and Good (1973) looking at
sediment yield in the Kosciusko area, while Blong *et al*
(1982) considered plot-scale erosion following a forest
wildfire near Narrabeen lagoon. Leitch *et al* (1983)
reported on the major Ash Wednesday fires in Victoria.

In several projects pre-fire monitoring had been
carried out. In the A.C.T. Bushrangers Experiment, the
catchment was deliberately burnt after a calibration period.
O'Loughlin *et al* (1982) examined discharge responses in
terms of increases in both baseflows which were due to
reduced transpiration and storm flows which were explained
by changes in contributing area. An uncontrolled wildfire
burnt most of the experimental catchments in the Eden
hydrology project after two years of record had been obtained
which also allowed a more detailed analysis of fire impacts.
Mackay *et al* (1980) and Mackay and Cornish (1982) examined
the changes in stream discharge, while Burgess *et al* (1981)
and Olive and Rieger (1987) reported on changes in suspended
sediment transport. In these studies the changes attributed
to fire were greater than those associated with logging.
Cornish and Binns (1988) found no significant change in
observed turbidity while Mackay and Robinson (1987) reported
similarly for stream chemistry.

III.   EDEN HYDROLOGY PROJECT

The Eden hydrology project is a paired catchment
experiment involving six small catchments in the Wallagaraugh
River in south eastern New South Wales. It was initiated
to examine the impact of clearfall logging operations
associated with the woodchip industry. A number of research
organisations have been involved with the N.S.W. Forestry
Commission examining basic hydrology and water quality: the
CSIRO studying run-off generation and hydrological response,
and the Soil Conservation Services overall erosion. Published
results of this work have been outlined in the preceding
section. This essay concentrates on the analysis of
suspended sediment transport carried out by the authors.

A.   General Environment and Experimental Design

The catchments and their locations are shown in Figure 1.   They vary in size from 75 to 225 ha, with elevations ranging from 180 to 476 m.   The underlying lithology is coarse-grained adamellites on which shallow red and yellow duplex soils have developed.   Average annual rainfall is approximately 900 mm with no marked seasonal pattern. However, inter- and intra-annual variation is high (Mackay and Robinson, 1987).   The resulting vegetation is dry sclerophyll eucalypt forest and severe intensity wildfires are a feature of this area.

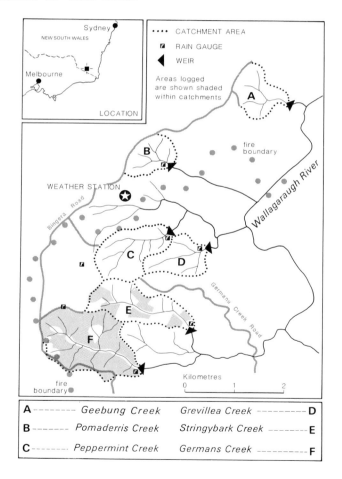

FIGURE 1.   Study area.

In all discharge, based on a 140° V-notch weir, and rainfall have been continuously monitored at the catchment outlet. Water samples have been collected on a routine weekly basis, while automatic samplers have provided more detailed storm sampling generally at hourly intervals. Suspended sediment concentrations have been determined by membrane filtration.

Using the paired catchment approach, monitoring began in 1977. After a short calibration period, Stringybark Creek catchment was logged in 1978 using the small alternate ·coupe system. By January 1979, 36% of the catchment area had been logged when an intense wildfire burnt four of the catchments in the research area (Fig. 1). Following this, Germans Creek, one of the burnt catchments, was clearfall salvage logged (86% of the catchment area). During 1987 two additional catchments have been logged using the alternate coupe technique.

B. Results

Rainfall during the study period varied considerably with annual totals ranging from 478 mm in 1980 to 1616 mm in 1978. Data from Eden, the nearest station with a long term record, indicated three years of above average rainfall (1978, 1983 and 1984) and four years of drought· from January 1979 to February 1982 (Moore *et al*, 1986). The rainfall pattern was dominated by relatively large storm events with one event in June 1978 yielding 339 mm of rainfall. Runoff reflects this rainfall variability and stream flow is very much dominated by storm events as shown in the annual hydrographs for Pomaderris Creek (Fig. 2). This hydrograph also shows the importance of large events, particularly those in mid 1978.

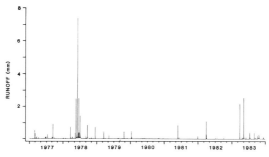

FIGURE 2. Pomaderris Creek annual hydrographs 1977-1983.

Suspended sediment concentrations in the undisturbed catchments were low with peak concentrations generally less than 100 mg L^{-1} and transport was storm dominated with sediment peaks of short duration, usually only a matter of hours. The response of sediment during storm events was highly variable and an analysis of 20 storms revealed six recognisable response patterns, while in 40% of the cases there was no recognisable pattern (Olive and Rieger, 1985). The generalised response patterns are shown in Figure 3. In multiple-rise storm events common in the larger events, there was clearly-defined sediment depletion or exhaustion (Fig. 4). In the initial rise suspended sediment concentrations were high, while in each succeeding rise peak concentration was lower until there was virtually no sediment response in later rises. The supply of sediment during storms is clearly limited and it decreases during the event.

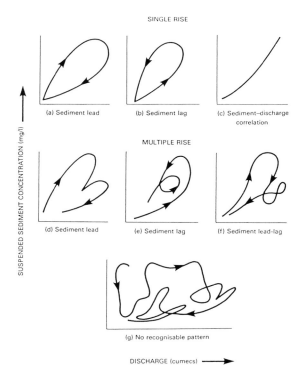

SINGLE RISE

(a) Sediment lead   (b) Sediment lag   (c) Sediment–discharge correlation

MULTIPLE RISE

(d) Sediment lead   (e) Sediment lag   (f) Sediment lead-lag

(g) No recognisable pattern

SUSPENDED SEDIMENT CONCENTRATION (mg/l)

DISCHARGE (cumecs) ⟶

FIGURE 3.   Generalised suspended sediment storm response patterns.

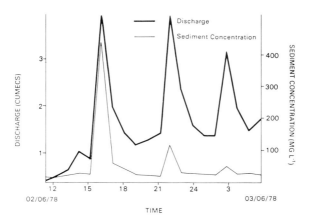

FIGURE 4. Sedigraph for Peppermint Creek for storm 2nd to 3rd June 1978.

Because of the variability of flow regimes and storm-sediment responses, quantification of the impact of catchment disturbance on suspended sediment is difficult and care must be taken in interpreting the results. In the logged Stringybark Creek, it appeared that sediment concentrations increased following logging with a maximum concentration of 793 mg L^{-1} occurring in the storm in June 1978. This was higher than peaks recorded in the catchment before logging. However, this event was much larger than those previously monitored. A comparison with the other catchments, where peak concentrations ranged from 160 to 400 mg L^{-1} for the same storm, indicated an increase in the logged catchment. Following the wildfire in January 1979 the burnt catchments had concentrations which appeared approximately two to three times higher than the pre-disturbance levels, while in the burnt and logged Stringybark Creek larger increases were apparent with a maximum concentration of 2520 mg L^{-1} in March 1979. In Germans Creek during salvage logging in October 1979, concentrations increased markedly with a peak storm concentration of 985 mg L^{-1}.

While there was an increase in the magnitude of sediment concentration, the storm-response patterns did not appear to change and in multiple-rise events sediment depletion was still clearly evident as is shown in Stringybark Creek during the storm in March 1978 (Fig. 5). This was despite the fact that all vegetative surface protection had been removed by the fire and the surface had been disturbed by logging operations leaving the potential for erosion at a maximum. Peak sediment concentrations were much higher but these were of short duration and, as the storm progresses, the sediment response pattern is one of reducing peaks with the supply of sediment becoming more limited.

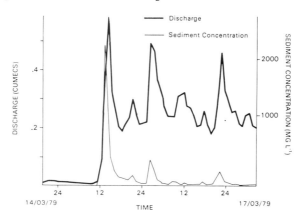

FIGURE 5. Sedigraph for Stringybark Creek from storm 14th to 17th March 1979.

Suspended sediment loads were only calculated for storm events which had been adequately sampled because of the errors involved in available estimation techniques (see Rieger and Olive, this volume). These were calculated using concentration data from automatic samples and the discharge series. In this study, sediment transport is completely dominated by two large storm events in May and June 1978 which accounted for up to 70-80% of the measured load from 1977 to 1983. Changes in loads due to disturbance have been reported previously and were summarised in Olive and Rieger (1987). A double-mass curve analysis revealed distinct breaks associated with the occurrence of logging and the wildfire (Fig. 6). The largest increases were apparently associated with the dual disturbances of logging and fire, while the latter appeared to have a greater impact than logging. Again quantification of the increases is difficult because of the variability in the streams.

The increases in suspended sediment loads during storms were generally less than expected, given the magnitude of the increases in peak sediment concentrations reported above, and stream discharge (Mackay and Cornish, 1982) which were up to ten fold in the streams suffering the dual disturbance of fire and logging. Load is a function of both sediment concentration and stream discharge and is determined by the

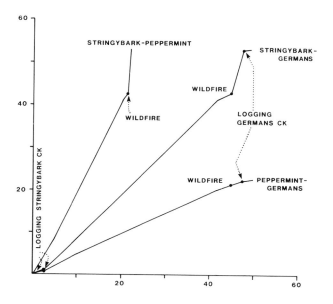

FIGURE 6.   *Suspended sediment doublemass curves (in tonnes) showing logging impact.*

general relationship

$$L = KCQ(t) \qquad\qquad (1)$$

where L is load in tonnes, C is sediment concentration, Q is stream discharge, t is a time function and K is a constant. Joint increases in concentration and discharge are multiplied to give larger increases in load, but this was not evident in this study and appeared to result from a number of other factors. The streams had a relatively rapid storm response before disturbance but this became more rapid after it. So while peak discharges and concentrations were increased, the maximum levels occur for only a short time. Coupled with this is the variability of storm-sediment response where there is commonly little correlation between the peaks of stream discharge and sediment concentration. This is particularly marked in the multiple rise events where sediment exhaustion is evident. There is a flush of sediment in the first stream rise where stream discharge may not be great while, in later larger stream rises, there is a subdued sediment response as is shown in Germans Creek in October 1979 following fire and salvage logging (Fig. 7). The result of combining these factors is that loads are much less than would be predicted using average discharge and sediment data from the storm.

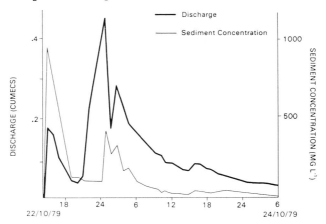

FIGURE 7. Sedigraph for Germans Creek for storm 22nd and 24th October 1979.

IV.   DISCUSSION

An important consideration to come out of the Eden research is the influence of the variability of rainfall on the hydrologic behaviour of the streams and the role this plays in determining the impact of any catchment disturbance. This also leads to questions about the validity of both the experimental design and analysis of the various disturbances.   Australian stream regimes appear more highly variable than for many other parts of the world and this is particularly marked with respect to the areas where many of the traditional techniques for analysis have been developed, notably in North America and Europe (see Finlayson and McMahon, this volume). Much of south-eastern Australia has no distinct seasonality of rainfall and stream-flow regimes are dominated by long periods of low flow interspersed with irregular storms of widely varying magnitude which can occur throughout the year.   In the case of suspended sediment, transport is commonly dominated by such stream events particularly those larger and less frequent.   In this situation, any analysis of the impact of catchment disturbance is made difficult because it is superimposed on an already-variable pattern.   This makes it difficult to distinguish confidently between the natural climatic variability and those changes due to disturbance.   The influence of this background variability is both temporal and spatial and operates at a number of different scales.

A. *Variability and Catchment Disturbance*

At the largest temporal scale the impact of any disturbance, be it a forestry operation or fire, is very much dependent on the rainfall regime which occurs associated with, and immediately following, the disturbance. If disturbance is associated with a period of below average rainfall or low intensity storm events, then sediment transport and erosion can be small in absolute terms. Where there are a series of low intensity storm events, these can result in relatively rapid regeneration of vegetation. This reduces the potential for erosion while not providing sufficient energy to carry out major erosion while the catchment is in the disturbed state. If however, disturbance corresponds with large storm events, then the resulting erosion can be substantial. This situation is clearly illustrated in the Eden research, where there has been a great deal of variability over the ten year study period. Relative increases were apparent in suspended sediment concentrations as a result of both logging and fire. However, the resulting sediment transport was very much dependent on the rainfall regime. The logging of Stringybark Creek during 1978 corresponded with a wet period and several large, high-intensity storms and so in absolute terms a large volume of material was transported. Following the fire there was a protracted drought and, while sediment concentrations were much higher, little erosion resulted. In the streams which were undisturbed before the fire, sediment transport in the ten year period is dominated by the large storms of 1978 and little material was transported following the fire.

At the storm scale, there can also be considerable variation in response associated with the magnitude and intensity of rainfall and the antecedent moisture conditions. This was particularly marked in the Eden catchments where there was considerable variation in suspended sediment response. This variation in response pattern is important in determining the quantity of sediment transported. With short-lived peaks of discharge and sediment concentration, and the sediment exhaustion through multi-peaked events, loads can vary considerably. So at this scale, sediment transport is very much dependent on the characteristics of individual storms.

B.   Variability and Experimental design

In the commonly-used paired-catchment approach, a number of problems become evident in its application to Australia. These relate to the variability of flow regimes. The first is in the amount of pre-disturbance calibration required to characterise adequately the behaviour of any stream. With the large variability, a very long record may be required and, because of the important role of individual storms, it may be that it is almost impossible to characterise confidently stream behaviour. This has been marked in the Eden research where sediment transport is related not only to the magnitude and nature of events, but also to the antecedent conditions and the temporal relationship of storms to each other. Because of the marked variability during the study, it is still difficult to predict storm-sediment response in the natural catchments.

Problems also arise with climatic variability during the disturbance period. Once the treatment programmes such as logging are initiated, the results are determined to a large extent by the rainfall regime during the disturbance period. The disturbance is often relatively short lived and may have a very distinctive rainfall pattern which is different to the pre-disturbance pattern. This was the case in the Eden study where the initial logging of Stringybark Creek was associated with a wet period with several very large storms, while in the period following the wildfire drought conditions prevailed with rainfalls well below average and very few storm events. Because of these major scale variations any valid comparison is difficult. The Eden results may indicate an extreme situation but illustrate the problems which are likely to emerge. In the paired-catchment approach, an attempt is made to overcome these problems by the use of control catchments. However, if the paired catchments differ in their response for a particular storm and between storms, as is the case in the Eden catchments, then the value of the control is severely reduced. It becomes impossible to distinguish normal variation from that due to catchment disturbance.

As well as the paired-catchment approach much of the previous research in Australia has used relatively simple statistical analysis of changes due to catchment disturbance. This has been particularly the case with respect to water-quality parameters including suspended sediment. A common technique has been to compare mean concentrations before and after disturbance (Langford et al,

1982). This type of analysis makes no allowance for
variation in streamflow regime over the period of the study,
where variations in the magnitude of storms could be more
important than any catchment disturbance. For example,
such an analysis of the impact of fire in the Eden catchments
would produce anomalous results, as pre-disturbance flows
were dominated by a number of large storm events yielding
high sediment concentrations. Following the fire, drought
conditions prevailed and, while concentrations were
relatively high, the absolute values were low because of the
small storms. While such techniques may be successfully
applied in environments where flow regimes are less variable,
their applicability to many Australian conditions is
doubtful.

The use of measures of central tendency, such as means
or medians, is also effectively applying a filter to the data
which tends to dampen out the extreme values (Burt, 1986).
In studies of catchment disturbance it is the extreme values
which are of greatest interest. However, because of their
less frequent occurrence, they are lost in the filtering
process. Again the influence of such filtering in using
these statistics is more important in a variable environment.

Attempts have been made to overcome the problems of
variability by comparing relative rather than absolute values
(Burgess *et al*, 1981; Cornish and Binns, 1988). While this
attempts to overcome the problems of the varying scale of
storm events, it is still insufficiently detailed to overcome the variability of response between storms. Mackay
and Cornish (1982), in assessing the impact of logging and
fire on storm discharge, have addressed these problems by
considering storm characteristics as well as magnitude to
establish general relationships. Such an analysis is less
quantitative but it does make more allowance for background
variability and produces more meaningful results. There
needs to be considerably more emphasis on understanding
underlying processes to examine changes and less on
statistical testing which makes no allowance for the underlying variability of the natural undisturbed system.

In studies of water quality or sediment transport,
temporal variability also needs to be taken into account when
designing the sampling strategy. Many studies are based on
a routine sampling with samples collected at regular
intervals (eg, weekly) (Cornish and Binns, 1988). They
reported on turbidity levels in the Eden catchments. In
the case of sediment, movement is dominated by storm events
and, as most studies have involved small catchments, these
are of relatively short duration. In the Eden catchments,

stream rises are rapid and usually only extend over a few hours and sediment responses are even more short lived (Figs. 2, 5 and 7). So the sampling strategy must be adequate to catch such shortlived events. The probability of sampling such occurrences with routine weekly sampling is very low. If a measure of central tendency is then used to summarise the data, the filtering process mentioned above reduces the influence of any extreme values sampled. Extreme values are important and therefore the research design and analysis techniques should not diminish their role.

An adequate sampling routine is also critical for determining sediment loads. Many calculations of changes in loads associated with disturbance are based on very limited data and involve estimation techniques which assume a single constant relationship between discharge and suspended sediment concentrations. In the Eden study, the relationship between these was highly variable, with peaks commonly out of phase and the response of sediment concentration changing through storms due to sediment depletion. As discussed in the results section, use of simple estimation techniques would substantially over estimate loads. A realistic load can only be determined by detailed sampling of storm events.

In such variable environments, a number of problems emerge from the use of the paired catchment approach and simple statistical analyses to assess the impact of catchment disturbance. This is especially important with respect to sediment transport. The underlying assumptions of such an approach are that the response patterns of the monitored variables are relatively constant through time and space, and that any changes are the result of catchment disturbance. In the Eden catchments this is clearly not the case and it is difficult to distinguish between variation due to disturbance and that related to background natural catchment variability. With this realisation there has been increasing emphasis in this and other research projects to focus research on more fundamental studies of the processes which operate in the system in an attempt to define more clearly changes due to disturbance. While superficially less quantitative, these are likely to provide more accurate characterisations of the impact of catchment disturbances.

V.   CONCLUSIONS

A number of problems emerge from the study of Australian forest catchment disturbance associated with the high variability in the natural undisturbed systems. This high background variation makes the assessment of the changes due to catchment disturbance difficult and raises a number of questions about the use of traditional analysis techniques developed in less variable environments. Given the high natural temporal and spatial variations of catchments, the use of the traditional paired catchment approach needs to be seriously questioned as the underlying assumption is that any change is due solely to catchment disturbance. Also the results obtained from the use of simple statistical techniques may be as much related to background variation as to disturbance. They may in fact filter out the most important extreme values.

The recent trend away from simple paired catchment experiments to more detailed studies of the underlying processes which operate in catchments and the changes in these processes which result from disturbance have come out of the realisation of the shortcomings of the application of traditional techniques.   An example of such research is reported by Bonell in this volume. In the case of sediment transport, there needs to be a thorough understanding of the sources of both runoff and sediment and the processes whereby the sediment is delivered to the fluvial system before the natural variability of the system can be explained.   Only when such processes are better understood can the impact of catchment disturbance be assessed with any confidence.

ACKNOLWEDGEMENTS

The authors would like to acknowledge the assistance of the Forestry Commission of New South Wales in the operation of the field work and the financial assistance provided by the Australian Research Grants Scheme.

ADDENDUM

A major project examining the impact of clearfall logging of Jarrah forest has been based on seven catchments near Manjimup (Water Authority of Western Australia, 1987).   Short term increases in water tables were reported which resulted in increases in stream salinity. Sediment concentrations also increased. The changes persisted for approximately two years and returned to pre-disturbance levels in four to five years.

REFERENCES

Blong, R.J., Riley, S.J. and Crozier, P.J. (1982) Sediment yield from runoff plots following bushire near Narrabeen Lagoon, N.S.W. *Search*, 13, 36-38.
Bren, L.J. and Leitch, C.J. (1985) Hydrologic effects of a stretch of forest road. *Aust. For. Res.*, 15, 183-194.
Bren, L.J. and Leitch, C.J. (1986) Rainfall and water yields of three small forested catchments in north-east Victoria, and relation to flow of local rivers. *Proc. Roy. Soc. Vic.*, 98, 19-29.
Brown, J.A.H. (1972) Hydrologic effects of bushfire in a catchment in south-eastern New South Wales. *J. Hydrol.*, 15, 77-96.
Burgess, J.S., Olive, L.J. and Rieger, W.A. (1981) Sediment yield change following logging and fire effects in dry sclerophyll forest in southern New South Wales. *IAHS Publ.*, 132, 375-385.
Burt, J.E. (1986) Time averages, climatic change and predictability, *Geog. Analysis*, 18, 279-294.
Cornish, P.C. (1980) *Karuah Hydrology Research Project, Annual Report 1979*, Forestry Commission of New South Wales.
Cornish, P.C. and Binns, D. (1988) Changes in streamflow quality following logging and wildfire in a dry sclerophyll forest in south eastern Australia. I. Turbidity. *For. Ecol. Manag.*, 22, 1-28.
Davies, T.R.H. and Pearce, A.J. (eds.) (1981) Erosion and sediment transport in Pacific rim steeplands. *IAHS Publ.* No. 132.
Douglas, I. (1973) Rates of denudation in selected small catchments in eastern Australia. *Univ. Hull Occasional Papers in Geography*, 21.
Gilmour, D.A. (1971) The effects of logging on streamflow and sedimentation in a north Queensland rainforest catchment. *Comm. Forestry Rev.*, , 50, 38-49.
Gilmour, D.A. (1977) The effects of logging and clearing on water yield and quality in a high rainfall zone of north east Queensland. *Inst. Engrs. Aust., Hydrol. Symp.*, Pub. No. 77/5, 155-160.

Gilmour, D.A., Cassells, D.S. and Bonell, M. (1982) Hydrology research in the tropical rainforests of North Queensland. Some implications for land management. *Inst. Engrs. Aust. Publ.*, No. 82/6, 145-152.

Good, R.B. (1973) A preliminary assessment of erosion following wildfires in Kosciusko National Park, NSW in 1973. *J. Soil Cons. NSW*, 29, 191-199.

Langford, K.J. and O'Shaughnessy, P.J. (1979) Second progress report North Maroondah. Melbourne and Metropolitan Board of Works, Melbourne.

Langford, K.J. and O'Shaughnessy, P.J. (1980) Second Progress Report - Coranderrk. Melbourne and Metropolitan Board of Works, Report No. MMBW-W-00104.

Langford, K.J., Moran, R.J. and O'Shaughnessy, P.J. (1982) The Coranderrk Experiment - the effects of roading and timber harvesting in a mature mountain ash forest on streamflow yield and quality. *Inst. Engrs. Aust. Publ.*, No. 82/6, 92-102.

Leitch, C.J. (1982) Sediment levels in tributaries of the East Kiewa River prior to logging alpine ash. *Inst. Engrs. Aust. Publ.*, No. 82/6, 72-78.

Leitch, C.J., Flinn, D.W. and van de Graft, R.J. (1983) Erosion and nutrient loss resulting from Ash Wednesday (16 February 1983), wildfires: a case study. *Aust. For.*, 46, 173-180.

Loughran, R.J. (1984) Studies of suspended sediment transport in Australian drainage basins - a review. In R.J. Loughran (ed.) *Drainage Basin Erosion and Sedimentation*, Univ. of Newcastle, NSW, 139-146.

Mackay, S.M. and Cornish, P.C. (1982) Effects of wildfire and logging on the hydrology of small catchments near Eden, NSW. *Inst. Engrs. Aust Publ.*, No. 82/6, 111-117.

Mackay, S.M. and Robinson, G. (1987) Effects of wildfire and logging on streamwater chemistry of small forested catchments in south eastern New South Wales. *Hydrol. Processes*, 11, 359-384.

Mackay, S.M., Mitchell, P.A. and Young, P.C. (1980) Hydrologic changes after wildfire in small catchments near Eden, NSW, Hydrology and Water Resources Symposium, Adelaide, *Inst. Engrs. Aust.*, 150-156.

Moore, I.D., Mackay, S.M., Wallbrink, P.J., Burch, G.J. and O'Loughlin, E.M. (1986) Hydrologic characteristics of a small forested catchment in south-eastern New South Wales. Part I Pre-logging condition. *J. Hydrol.*, 83, 307-335.

Olive, L.J. and Rieger, W.A. (1985) Variation in suspended sediment concentration during storms in five small catchments in south east New South Wales. *Aust. Geog. Studies*, 23, 38-51.

Olive, L.J. and Rieger, W.A. (1987) Eden catchment project: sediment transport and catchment disturbance, 1977-1983. Monograph Series No. 1, Dept. Geography and Oceanography, ADFA, Canberra.

Olive, L.J. and Walker, P.H. (1982) Processes in overland flow - erosion and production of suspended material. In E.M. O'Loughlin and P. Cullen (eds.) *Prediction in Water Quality*, Aust. Acad. Sci., Canberra, 87-120.

O'Loughlin, C.J. and Pearce, A.J. (eds.) (1984) Symposium on effects of forest land use on erosion and slope stability. East-West Center, University of Hawaii, Honolulu.

O'Loughlin, E.M., Cheney, N.P. and Burns, J. (1982) The Bushrangers Experiment: hydrologic response of a eucalypt catchment to fire. *Inst. Engrs. Aust. Publ.*, No. 82/6, 132-138.

Rieger, W.A. and Olive, L.J. (1986) Sediment responses during storm events in small forested catchments. In A.H.El. Shaarawi and R.E. Kwaitkowski (eds.) *Statistical Aspects of Water Quality Monitoring*, Developments in Water Sciences 27, Elsevier, Amsterdam, 390-398.

Rieger, W.A., Olive, L.J. and Burgess, J.S. (1979) Sediment discharge response to clear-fell logging in selected small catchments, Eden, NSW. *Proc. 10th N.Z. Geog. Conf.*, Auckland, 44-48.

Riley, S.J. (1984) Effect of clearing and road operations on the permeability of forest soils, Karuah Catchment, NSW, Australia. *For. Ecol. Manag.*, 9, 283-293.

Sharma, M.L., Johnston, C.D. and Barron, R.J.W. (1982) Soil water and groundwater responses to forest clearing in a paired catchment study in south-western Australia. *Inst. Engrs. Aust. Publ.*, No. 82/6, 118-123.

Stokes, R.A. and Loh, I.C. (1982) Streamflow and solute characteristics of a forested and deforested catchment pair in south-western Australia. *Inst. Engrs. Aust. Publ.*, No. 82/6, 60-66.

Water Authority of Western Australia (1987) *The Impact of Logging on the Water Resources of the Southern Forests, Western Australia*, Report by the Steering Committee for Research on Land Use and Water Supply, Report No. WH 41, 33p.

15
Fluvial Dispersion of Radioactive Mill Tailings in the Seasonally-Wet Tropics, Northern Australia

T.J. East
R.F. Cull

Alligator Rivers Region Research Institute
Jabiru East, NT

A.S. Murray

CSIRO
Division of Water Resources Research
Canberra, ACT

K. Duggan

N.T.T.I.A.D. Project
Kupang NTT, Indonesia

I. INTRODUCTION

     Mining in the seasonally wet tropics of North Australia
poses special environmental problems, associated mainly
with the marked seasonality of high intensity (summer)
rains especially at the onset of the Wet when any vegetation
cover is most depleted (Duggan, 1985; Applegate *et al*,
1986). A major expansion of mineral exploitation at the
end of World War II caused significant environmental impacts.
This coincided with the development of large scale open-
cut mining techniques. More recently, there has emerged
a growing awareness in the community of environmental
values. At the Rum Jungle Mine in the Northern Territory

(Fig. 1), where uranium and copper ores were mined between 1952 and 1971, containment of mine wastes was ineffectual and consequently more than 100 km^2 of the Finniss River floodplain were contaminated with radioactive tailings, pyritic acid leachates and heavy metals (Davy, 1975). This resulted in the destruction of aquatic fauna and riparian vegetation, the latter leading to accelerated bank erosion and channel siltation. This mine was eventually rehabilitated by the Australian Government between 1982 and 1987 at a public cost of over $20 mill. Similar but smaller scale patterns of environmental damage accompanied many mining operations in Northern Australia in this period.

The Rum Jungle experience highlighted the need for environmental protection and this essay is concerned with this in the present development of mining in the Alligator Rivers Region. An analogue is used to illustrate geomorphic processes and environmental degradation at Moline. This is helping to formulate rehabilitation policies for Ranger.

II. ENVIRONMENTAL PROTECTION AT URANIUM MINES IN THE ALLIGATOR RIVERS REGION

The Ranger Uranium Mine is located 250 km east of Darwin in the Alligator Rivers Region (ARR) (Fig. 1). It is situated within the Kakadu National Park listed by UNESCO on the World Heritage Register. Here the main feature of the environmental safeguards is the containment of the tailings dam, contaminated water retention ponds, pit, mill and ore stockpiles and all site runoff water within a bunded area - the Restricted Release Zone (RRZ). No release of water from this area has been permitted during seven years of operation. Stringent environmental requirements relate also to the rehabilitation of the decommissioned site and structures (Supervising Scientist, 1984-85), the most important of which will be the disposal of about 18 mill t of radioactive tailings soon after 1992. There are two options for disposal - burial in the mine pit or containment in the tailings dam (East, 1986; Dames and Moore, 1987). The erosion potential of engineered slopes here is high, and standards for rehabilitated structures are currently being formulated to ensure protection of residents and ecosystems downstream of the mine. It is almost inevitable that rehabilitation will be accompanied by some increase in sediment. In order to formulate standards for site rehabilitation, it is necessary to predict the sediment

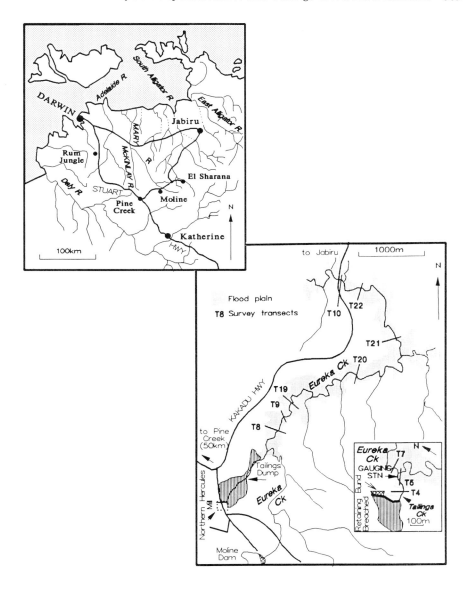

FIGURE 1. Location of Northern Hercules Mine, Moline and survey transects.

transport rates and depositional sites for mine wastes.   One
approach to this problem is to use an analogue site.

III. THE MOLINE STUDY

The Northern Hercules Mine at Moline, abandoned in 1972,
is an analogue for the study of fluvial transport of
eroded mine materials.   Earth bunds were constructed to
settle tailings, but not as long term retention structures.
Since the end of milling these have been eroded and deposited
on downstream floodplains (Pickup *et al,* 1987; Cull *et al,*
1986).   Available data on this environment indicate a broad
correspondence between the Moline catchment and those at
Ranger (East *et al,* 1987b).   The main aims in this study
were: (a) to measure flow and sediment transport processes
in Tailings Creek, and (b) to identify the patterns of
tailings deposition downstream of the eroding tailings dump
- in particular, variations in the radiological and size
characteristics of mine sediments within and between downstream sedimentary environments.   This was to provide an
analogue for predicting the possible fluvial dispersion of
mine sediments, at existing and future uranium mines in
this region.

A.   Tailings Production at Moline

The site is located about 50 km east of Pine Creek (Fig.
1).   The mill is situated at the head of Tailings Creek
which flows in turn into Eureka Creek, Bowerbird Creek and
finally into the Mary River, some 20 km from the mill.
Small scale operations began in the 1890s but the main
period of mining was from 1959 to 1972 when milling of
uranium and metal ores produced some 246,000 t of tailings.
In the mining period there were no controls for long-term
tailings containment.   The slurry was pumped onto the
natural ground surface and contained by earth bunds, the
last of which failed soon after 1972 (Fig. 2).   By 1982,
the estimated erosion was 63,000 t or about 25% of the
tailings (NTDME, 1982).   Sub-bankfull flows probably moved
most of the tailings with suspended flow load flushed through
the Mary River catchment into Van Diemen Gulf.   Overbank
flows, however, have deposited tailings on downstream
floodplains.
When dumped about 50% of the tailings were as silt and
clay (<63 µm) and the remainder was sand (Murray and Fisher,

1968). Based on a total of 130,420 t of uranium ore processed (Fisher, 1968), the average Ra-226 activity of the tailings is 34,500 Bq kg^{-1} (cf. Ra-226 activity of non-mineralised sediments of 50-100 Bq kg^{-1}).

B. Methods

Pickup *et al.* (1987) have shown that tailings deposited at Moline can be distinguished from the natural sediments by higher concentrations of heavy metals and CaCO$_3$. The radioactivity of the tailings can also be used as a 'label' to distinguish them from other sediments. This can be rapidly determined in the field, in spite of the mixing of both radioactive and nonradioactive tailings because levels are still greater than background.

An automatic stream gauge and water sampling station was established prior to the 1984/85 Wet Season on Tailings Creek, 200 m downstream of the tailings dump and 30 m from the confluence with Eureka Creek (Fig. 1). Flow in Tailings Creek was recorded and depth-integrated water samples were collected over the 1984/85 wet seasons. Water samples were analysed for total suspended solids by filtering through a 0.45 µm filter with residues being retained for radiometric analyses.

Field radiological surveys were carried out to determine tailings distribution between the lowest tailings dump bund and the Kakadu Highway crossing of Bowerbird Creek, 7.6 km downstream (Fig. 1). Ten transects across the floodplain were surveyed, and total gamma dose rates were measured at the ground surface with a gamma dose rate meter. Measurements were taken in different depositional environments at intervals not exceeding 25 m. These were categorised at the time of measurement as: slackwater sites, swales, levees, channels, backslopes, backswamps and valley sides. The floodplain stratigraphy (tailings and underlying sediments) was described from exposures in the banks, and in pits excavated on the floodplains of Tailings and Eureka Creeks (Fig. 3).

Laboratory radionuclide analyses of suspended solids were undertaken using high resolution gamma spectrometry (Murray *et al*, 1987). Suspended solids samples were ashed at 400°C to remove the filter paper and then cast in polyester resin to retain Rn-222, a gaseous daughter of Ra-226; sample size was between 0.2g and 8g. Typical minimum detection limits for the two radionuclides of

FIGURE 2.   *Moline tailings dump and abandoned mill with breached bund wall and active gullying.*

FIGURE 3.   *Bank of Tailings Creek 100 m downstream of lowest bund.*

interest, Ra-226 and Pb-210 are 2 Bq kg^{-1} and 15 Bq kg^{-1}
respectively.

C. Flow and Sediment Transport in Tailings Creek

The catchment area of Tailings Creek is about 75 ha and
about 22 percent is covered by the tailings pile. This creek
has an ephemeral flow regime. Peak discharges for eight
flows (or 43% of total discharge) sampled in the 1984/85
wet season ranged from 0.4 to 1.43 m^3 s^{-1} with flow durations
of between 2 and 20 hours. High flow levels in Eureka Creek
created some backflow problems at the Tailings Creek gauge
(Fig. 4a). This created some problems in establishing
stage-discharge relationships.

The dominant sediment source in this catchment since
1972 has been the eroding tailings pile. This pile has
since been devoid of vegetation with erosion by gullying
and rilling with sheet erosion in inter-rill areas (Fig.
2). During the 1984/85 wet season, concentrations of
suspended sediment varied between 10 and 94,000 mg L^{-1}, and
were greater than 1000 mg L^{-1} for all sampled flows. Peak
concentrations of suspended sediment generally occurred on
rising stages. However back flows caused decreased
sediment concentrations at high stages (Fig. 4a). On
falling stages when the back-flow effect was mitigated,
remobilisation of deposited sediment elevated concentrations
slightly (Fig. 5). Large individual floods were
characterised by high sediment yields. The maximum sediment
yield of 253 t for a single flow early in the 1984/85 wet
season (15/2/1985) accounted for over 50 percent of the
sampled annual yield. Cyclone Gretel produced two flows
in mid-April which represented a further 29 percent (144 t)
of the total yield of 504 t. The suspended yield for this
catchment was estimated from the stage-sediment concentration relationship (Fig. 6) because the flow-con-
centration relationship could not be accurately established.
As the gauging site in the wet season was accessible only
by helicopter, it was not possible to sample all flows for
suspended sediment, but a continuous hydrograph was
obtained.

The estimated 1984/85 wet season yield of 1200 t ($\pm$ 100)
represents about 0.7 percent of the estimated remaining
tailings mass. This is a mean denudation rate of 1.0 mm y^{-1}
over the total catchment area. If the rate for the
remaining 78 percent approximates the 0.07 mm y^{-1} obtained
by Williams (1973) for slopes in the Pine Creek area and
the 0.02 mm y^{-1} by East *et al* (1987a) for Ranger Mine area

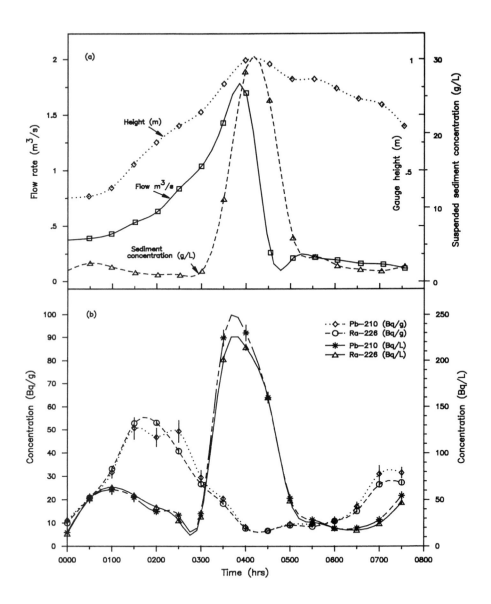

FIGURE 4. (a) Flow rate, gauge height and suspended sediment concentrations at Tailings Creek gauging station for Cyclone Gretel, 17/4/1985. Eureka Creek backflow reduces flow discharge at high stages; (b) Corresponding Ra-226 and Pb-210 characteristics for suspended solids.

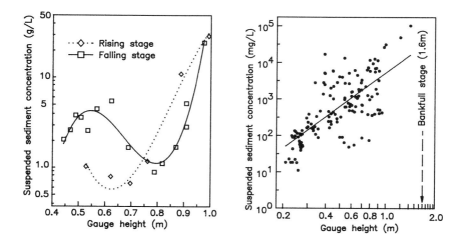

*FIGURE 5.   Relations between suspended sediment con-
centration and gauge height for Tailings Creek, showing
reduced concentrations at high (falling) stages because
of backflow effects (event no. 6, 17/4/85).*

*FIGURE 6. Suspended sediment concentration/gauge
height relations for Tailings Creek: all sampled events,
1984/85 wet season.*

catchments, then the mean erosion rate for the tailings area
is close to 4 mm y^{-1}, almost two orders of magnitude higher
than natural rates in the Pine Creek and Jabiru areas.

These 1200 t do not include bed load. It is difficult
to compare with the estimated total loss of 63,000 t from
the tailings pile. However, from grinding specifications
(50% < 75 μm) it is possible to estimate the total amount
of eroded material likely to have been available for
transport as suspended sediment. Of the 32,000 t coarser
than 75 μm, a large fraction would have been transported
as suspended solids; this is assumed to be 0.5 ± 0.25
(of the >75 μm material). This in turn implies that the
total transported as suspended load was 48,000 t (±8000)
giving a mean erosion rate of 4000 t (± 1700) y^{-1} for the 12
year period since 1972, compared with the 1984/85 figure
of 1200 t y^{-1}. Clearly there has been a marked reduction
in erosion rates on the tailings pile since closure (in spite
of 1984/85 being 12% wetter than average). Rates of gully
growth would have been highest just after closure, when
construction of the tailings pile had resulted in a local
oversteepening of the valley floor. Assuming that the
decrease in erosion rates since 1972 has been approximately

exponential and that a total of 48,000 t (±8000) has been
lost since then, it can be calculated that the suspended
sediment yield for 1972 would have been about 9000 t
(±2000) and that it took about 4±.6 y for the erosion rate
to drop by 50%. This is compatible with the expected gully
stabilisation whereby gully downcutting and headward growth
over time result in local base levels being approached with
a decline in associated sediment yields (cf. Tuan, 1966).

D. Radiometric Characteristics of Transported Material

The radiometric characteristics (Ra-226 and Pb-210
concentration) of the suspended sediment were analysed for
the first stage rise of Cyclone Gretel (17 April 1985)
(Fig. 4b). The specific activity rises to a peak of about 55
Bq g^{-1} during the low flows preceeding the floodpeak, drops
to about 6 Bq g^{-1} during the peak, and then rises again to
about 30 Bq g^{-1}. The high activities in low flows may be
explained, at least in part, by the affinity of radionuclides
for fine particulates whereby heavy metal ions including the
radioactive elements, are adsorbed preferentially onto
smaller grain sizes. This is particularly the case for
humic and clay colloids (<2 μm) (Scott, 1968; Gibbs, 1973;
Lewin *et al*, 1977; Kvasnicka, 1986; Marcus, 1987). Thus
variations during the hydrograph are likely to reflect
variations in size distribution of suspended solids
transported by different flow velocities and discharges.
Only the clay, silt and some fine sand would be in suspension
at low flows. These fractions would have greater specific
activities than the larger sized particles carried
additionally in suspension during the peak of the flood.
A comparison of the particle size characteristics of
suspended solids from high (1.2 m^3 s^{-1}) and relatively low
(<0.1 m^3 s^{-1}) flows, shows that sediment transported by high
flows is dominated by medium and coarse sand; in low flows,
particles coarser than fine sand (180 μm) were absent.
Exhaustion of the supply of fine particles at high discharges
may also account for an increase in the proportion of
coarser suspended material. The high activity of the fine
suspended material has implications for the possible contamination of downstream floodplains because it is only
the wash load (solutes and suspended solids) component that
is available for overbank deposition. Consequently,highest
radioactivity sediments would be deposited by vertical
accretion on floodplains, while coarse, low activity material
is contained largely within the channel.

E. Tailings Deposition and Floodplain Stratigraphy

The stratigraphy of Tailings and Eureka Creeks includes tailings varying between a few millimetres and 2.0 m capping the pre-mining floodplain sediments (Fig. 3). Multiple, thin, discontinuous layers of tailings are clearly distinguished by colour and in some cases particle size, reflecting an episodic depositional history.

Spot measurements indicated that gamma dose rates were significantly higher than (natural) background levels. In order to determine more precisely the radioactivity of the floodplain sediments and hence the distribution of tailings, ten transects were surveyed between the bund and the Kakadu Highway crossing (Fig. 1). The gamma dose rate profiles for selected transects (T7, T19 and T20) and corresponding topographic profiles are shown in Figure 7. Depositional environments included channels, levees, backslopes, swales, backswamps and slackwater sites. Dose rates were also recorded at the base of adjoining valley-side slopes.

Dose rates in seasonally flooded parts of the channel and floodplain were in the range 30-2000nGy h^{-1} (including instrument and cosmic ray background). These are consistently higher than those measured on similar topography in Eureka Creek upstream of the Tailings Creek confluence where the background is about 20nGy h^{-1}.

Dose rates generally decrease with distance from the tailings pile, although slackwater deposits are local anomalies. The average dose rate for each transect was calculated by weighting each measurement by the width of transect the measurement represents, and subtracting the natural background component of 20nGy h^{-1}. These have been plotted against the channel bed distance downstream from the lowest bund (Fig. 8). Two distinct trends are apparent. The first shows a high rate of decrease in mean dose rate from the bund to T7 just downstream of the Eureka Creek confluence. The second is characterised by a much lower rate of decrease in mean dose rate between T7 and T10. In the latter case the decrease is approximately exponential, similar to changes in a polluted Welsh river noted by Wolfenden and Lewin (1978). At Moline, the inflection point between the two segments (Fig. 8) is about 100 metres downstream of the confluence with Eureka Creek. The upstream segment consists of a continuous layer of poorly sorted, coarse tailings up to 2 m in thickness, extending across the floodplain and valley floor. The large quantity, poor sorting and high proportion of gravel in this material are indicative of mudflow processes which might have operated when the saturated tailings breached the bund (Fig. 2).

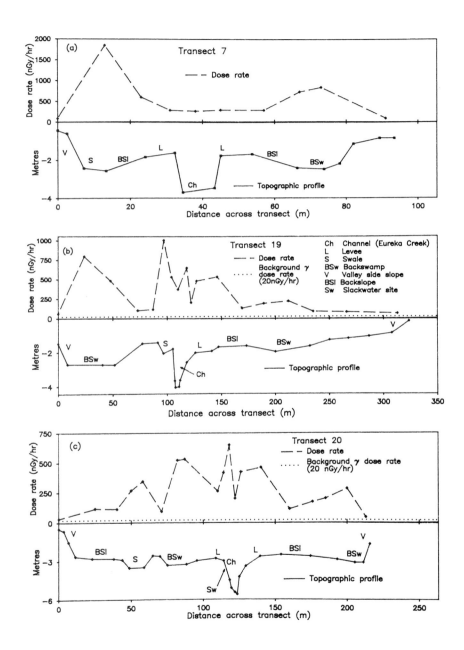

FIGURE 7.   Total gamma dose rate and topographic profiles for transects T7, T19 and T20 downstream of Moline.

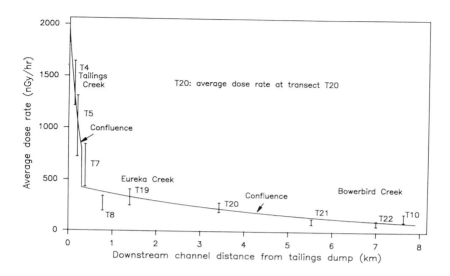

FIGURE 8. Profile of average gamma dose rates for floodplain downstream of Moline.

Mudflows transport sediment at extremely high concentrations over short distances, thereby inhibiting any significant size sorting (Pe and Piper, 1975). Downstream of this, transport is determined by fluvial processes in a distinct channel with the load separated into bedload, suspended load and solute components. It is these contrasting processes which may account for some of the reduction in the overbank deposition of tailings.

Generally consistent relations exist between the depositional environment and dose rate (Fig. 7; Table I). A one-way analysis of variance of gamma dose rates was used to identify statistically significant differences between the seven depositional environments. Prior to analysis, the individual recorded dose rates were normalised by dividing by the smoothed dose rate at the appropriate distance downstream, as calculated from the fitted exponential (Fig. 8). These normalised rates were further subjected to $\log_{10}$ transformation to establish homogeneity of variance. Only one sample (floodplain, no. 92; dose rate 6526 counts/ks) was greater than two standard deviations from the mean for that group and was omitted from subsequent statistical analysis. A one-way analysis of variance of dose rates showed statistically significant differences between the seven environments ($F=20.49$;

TABLE I.   *Mean gamma dose rates and tailings thickness for floodplain depositional environments.*

| | Depositional Environments | | | | | | |
	Slackwater	Swales	Levees	Channels	Backswamps	Backslopes	Valley Sides
No. of sites	8	18	26	11	28	33 [32][4]	10
Mean dose rate nGy/hr	900	640	360	340	215	160 [145]	40
S.E.nGy/hr	125	90	40	50	30	13 [11]	3
Tailings[1]. median visible depth, cm	28(6)[2]	13(16)	5(21)[3]	NR	0(25)	0(28)	0(10)

[1].   The presence of small amounts of tailings was often not detectable by eye, only by dose rate meter.

[2].   Number of sites at which observations were recorded.

[3].   Four levee sites had deposits of 11-18cm of sandy, non-radioactive sediment overlying the tailings.

[4].   Sample No. 92 omitted.

NR: No observations recorded.

df=6; p=0.0001).   These differences were further investigated using the Waller-Duncan K-ratio test and Duncan's multiple range test.   Both analyses produced essentially identical results with significant differences in mean dose rates characterising most depositional environments (Table II). These variations are explained in terms of sediment deposition processes and the particle-size properties of the (tailings) sediment on the floodplain.

TABLE II.   *Analysis of variance of mean dose rates between depositional environments: Summary of differences.*

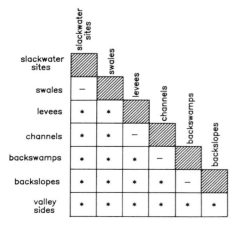

1. Differences marked with an asterisk (*) are significant at the 0.05 level; (−) indicates no significant difference.

   The highest dose rates were recorded by slackwater sites
($\bar{X}$ = 900mGy h^{-1}; n=8) where localised reductions in high flow
velocities allowed tailings accumulations up to 1.6 m thick.
At transect T10, a causeway created artificial slackwater
conditions with the deposition of nearly 0.7 m of tailings.
Gamma spectrometry analysis of a core from this site gave
a Ra-226 activity of 12500 Bq kg^{-1} at a depth of 0.5 m (Cull
et al, 1986). The size characteristics of the slackwater
deposits showed the weight percent of fines (silt and clay,
75%) and sand (25%) to be intermediate in value in relation
to the other depositional environments (Fig. 9). The
dominance of the fines reflects the low energy conditions
which are characteristic of slackwater sites (Kochel and
Baker, 1982) whilst the presence of some sand indicates
channel proximity. Slackwater mean dose rates were
significantly different to those for all environments except
for swales (Table I; Fig. 2).

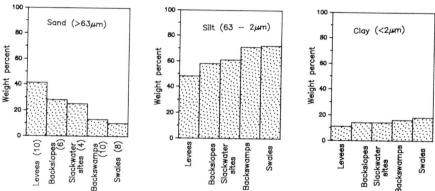

FIGURE 9. *Textural characteristics for surface samples
from floodplain sedimentary environments, Moline.*

   Swales include localised floodplain depressions
(meander cut-offs, chutes, flood by-pass channels) and have
tailings up to 70 cm thick (mean depth 13cm; n=16). Swale
deposits are characterised by the highest concentrations
of fines (Fig. 9, mean silt and clay, 90%; n=8). This is
consistent with the low overbank flow velocities and with
typically fine overbank deposition (Allen, 1965). Some
depressions such as chutes transmit excess flows but at lower
flow velocities. This may account for the comparatively
high deposition rates at these sites. The combination of
generally thick sequences of tailings and very high silt
and clay contents gave rise to high dose rates (mean, 640n
Gy h^{-1}; n=18) which are significantly different to all
other depositional environments other than slackwater sites

(Table II). As noted above, the fine fraction again appears
to be associated with the highest specific activities. This
is consistent with recorded instances of fine grained
sediments with high heavy metal concentrations being
deposited selectively in low energy sedimentary environments
(Davies and Lewin, 1974; Wolfenden and Lewin, 1977). A
similar pattern has been observed for uncontaminated
floodplains (Finlayson *et al*, 1984).

Generally, the lowest dose rates were associated with
elevated and less frequently flooded sites. Valley-side
basal slopes had mean dose rates of 40nGy h^{-1} (n=10) about
twice background levels. Backslopes behind levees recorded
higher dose rates (mean, 160nGy h^{-1}; n=33) but still low by
comparison with swales, slackwater sites and levees (Table
I). Backslopes generally terminated in backswamps where
dose rates (mean, 215nGy h^{-1}; n=28) were relatively low
because of the combined effects of local uncontaminated
drainage (fills the backswamps during the early wet season
before significant overbank flooding from the main channel)
and distance from the main channel. Rates are still an
order of magnitude above background rates. Thick riparian
vegetation on levees and wet season floodplain grasses
filter suspended solids in overbank flows. This results
in only the finest wash-load fraction being available for
backswamp deposition. At Moline, there is a systematic
increase in fines (silt and clay) away from the channel -
from levees through backslopes to swales and backswamps
(Fig. 10). There is also evidence for lower deposition rates
away from the channel (Table I). The results of the
analysis of variance of mean dose rates (Table II) showed
the backslopes to be the only class of depositional
environments not to be significantly different from the
backswamps.

Dose rates on the levees were intermediate (mean,
360nGy h^{-1}; n=26) between the more elevated and/or distant
parts of the floodplain and the topographic depressions
(swales). Mean dose rates for levees were significantly
different to all others except for the channel (Table II).
The thickness of tailings was highly variable (0cm to 39cm;
mean 5cm) but generally displayed fine, well-stratified
bedding readily distinguished by variations in colour
and particle size. Levee tailings were the coarsest of overbank deposits (Fig. 9) comprising 41 percent sand on
average; this is consistent with natural levee deposits.
This is a consequence of the proximity at the levees to
the main channel, and flow velocities intermediate between
those of the channel and backwater areas. Deposition

would be promoted by the reduction of flow velocities by bank vegetation. Surface dose rates were similarly intermediate in value reflecting the deficiency of fines. Differences in metal concentrations in low energy fine deposits (river cut-offs) and high energy coarse deposits (point bars) depositional environments have been observed in polluted Welsh streams by Wolfenden and Lewin (1977). However, no variability could be detected between the wider range of topographic features as reported here. This successful differentiation of mine sediments in a range of sedimentary environments has been possible here because of (a) high sediment inputs from a point source over a relatively short time span, (b) low natural sediment concentrations which meant that most of the load was tailings, (c) a range of generally well developed and readily recognisable sedimentary environments, and (d) the distinctive radiological characteristics of different sized sediment particles.

Because the floodplains of these meandering streams are continually reconstructed by in-channel erosion and deposition, as well as overbank deposition, the contaminated sediments will survive to be monitored in the future, even when the primary source of the tailings has gone (cf. Wolfenden and Lewin, 1977). The long half-life (8×10^4y) of the long-lived parent Th-230 means that the contaminated floodplain will remain more radioactive than the natural sediments for thousands of years.

IV. CONCLUSIONS

Erosion of tailings at a uranium mine abandoned in 1972 has resulted in large present-day inputs (up to 94g L^{-1}) of radioactive sediment into local watercourses after the failure of containment bunds. Mean erosion rates of the pile are now about two orders of magnitude higher than those for natural slopes in the region, but they have been decreasing rapidly as erosion systems stabilize. The fine (<63 μm) fraction of the suspended load is characterised by high specific activities, and it is this part of the load which is available for overbank deposition.

The downstream dispersal patterns of radioactive tailings are controlled by (a) the nature of sedimentary environments, (b) the properties of the tailings sediment which affect transport, and (c) the dilution of flow and sediment from incoming tributaries. A generally consistent relationship exists between the type of sedimentary floodplain environment and the surface gamma dose rate. These differences in mean dose rate between classes of depositional environment were shown to be statistically significant in the majority of cases. The low energy environments of swales and slackwater sites are characterised by the thickest sequences of tailings and the highest dose rates. Low deposition rates of high activity fine tailings on blackslopes and in adjoining back-swamps resulted in comparatively low mean dose rates but they are lowest on the valley-side slopes. Generally coarse but thick, overbank deposits on levees exhibited intermediate dose rates. These patterns reflect the size fractionation of tailings during fluvial transport and overbank flow, with the deposition of different sized sediments in the various sedimentary environments and the tendency for fine particles to be more radioactive.

Dose rates generally decrease with distance downstream from source. The rate of decrease is not constant, with high rates of decrease in immediate catchment, changing abruptly to lower rates of decrease below the confluence with Eureka Creek.

ACKNOWLEDGEMENTS

The authors wish to thank John MacCartie for his very considerable assistance in the field. The contributions of Ms Andrienne Nankivell, Ms Michelle Templeman, John Pfitzner and Mrs Patricia McCallum in the preparation of figures and in typing the manuscript are gratefully acknowledged.

REFERENCES

Allen, J.R.L. (1965) A review of the origin and characteristics of recent alluvial sediments. *Sedimentol.*, 5, 89-191.

Applegate, R.J., Burgess, J.W., Duggan, K. and Tatzenko, S.P. (1986) Erosion assessment report on Stage 1 of Kakadu National Park. *Land Cons. Unit, Cons. Comm. N.T. Tech. Rept.*, No. 24, Darwin.

Cull, R.F., Duggan, K., East, T.J., Martin, R. and Murray, A.S. (1986) Tailings transport and deposition downstream of the Northern Hercules (Moline) Mine in the catchment of the Mary River, N.T. In P.J.R. Broese Von Groenou and J.R. Burton (eds.), Environmental Planning and Management for Mining and Energy. *Proc. N. Aust. Mine Rehabil. Workshop No. 10,* Darwin, 199-216.

Dames and Moore (1987) *Conceptual Planning for Rehabilitation of the Ranger Project Area.* Report for Ranger Uranium Mines.

Davies, B.E. and Lewin, J. (1974) Chronosequences in alluvial soils with special reference to historic lead pollution in Cardiganshire, Wales. *Environ. Pollution,* 6, 49-57.

Duggan, K. (1985) Erosion and sediment transport in the lowlands of the Alligator Rivers Region, N.T. In K.N. Bardsley, J.P.S. Davie and C.D. Woodroffe (eds.) Coasts and Tidal Wetlands of the Australian Monsoon Region. *Mangrove Monograph No. 1,* North Aust. Res. Unit, ANU, 53-63.

Davy, D.R. (ed.) (1975) Rum Jungle Environmental Studies. *Aust. Atom. Energy Comm.,* Rept. AAEC/E365.

East, T.J. (1986) Geomorphological assessment of sites and impoundments for the long term containment of uranium mill tailings in the Alligator Rivers Region. *Aust. Geog.,* 17, 16-21.

East, T.J., Cull, R.F. and Duggan, K. (1987a) Surface runoff yields and sediment transport processes near Ranger. *Alligator Rivers Region Research Institute (ARRRI) Res. Summ. 1985-1986,* 17-20.

East, T.J., Cull, R.F., Murray, A.S., Marten, R. and Duggan, K. (1987b) Tailings erosion and transport at Moline. *ARRRI Res. Summ. 1986-1987,* 21-24.

Finlayson, M., Johnston, A., Murray, A.S., Marten, R. and Martin, P. (1984) Radionuclide distribution in sediments and macrophytes. *ARRRI Res. Rept. for 1983-1984,* 37-42.

Fisher, W.J. (1968) Mining practice in the South Alligator Valley. In D.A. Berkman, R.H. Cuthbert and J.A. Harris (eds.) *Proceedings of the Symposium on 'Uranium in Australia",* Aust. Inst. Min. Metall., Rum Jungle Branch.

Gibbs, R.J. (1973) Mechanisms of trace metal transport in rivers. *Science,* 180, 71-73.

Kochel, R.C. and Baker, V.R. (1982) Palaeoflood hydrology. *Science,* 215, 353-361.

Kvasnicka, J. (1986) Radiation data input for the design of dry or semi-dry uranium tailings disposal. *Health Physics,* 51, 329-336.

Lewin, J., Davies, B.E. and Wolfenden,P.J. (1977) Inter-actions between channel change and historic mining sediments. In K.J. Gregory (ed.) *River Channel Changes,* Wiley, Chichester, 353-367.

Marcus, W.A. (1987) Copper dispersion in ephemeral stream sediments. *Earth Surf. Proc. and Landforms*, 12, 217-228.

Murray, A.S., Marten, R., Johnston, A. and Martin, P. (1987) Analyses for naturally occurring radionuclides at environmental concentrations by gamma spectrometry. *J. Radioanal. and Nuclear Chem.*, 115, 263-288.

Murray, R.J. and Fisher, W.J. (1968) The treatment of the South Alligator Valley uranium ores. In D.A. Berkman, R.H. Cuthbert and J.A. Harris (eds.) *Proceedings of the Symposium on 'Uranium in Australia"*, Aust. Inst. Min. Metall., Rum Jungle Branch.

Northern Territory Department of Mines and Energy (NTDME) (1982) *Moline and South Alligator Mills Tailings Dumps Site Investigations for Potential Rehabilitation*, Environ. Prot. Branch, Darwin.

Pe, G.G. and Piper, D.J.W. (1975) Textural recognition of mudflow deposits. *Sed. Geol.*, 13, 303-306.

Pickup, G., Wasson, R.J., Warner, R.F., Tongway, D. and Clark, R.L. (1987) A study of geomorphic research for the long term management of uranium mill tailings. *CSIRO Div. Water Resourc. Res. Div. Rept. 87/2*, Canberra.

Scott, M.R. (1968) Thorium and uranium concentrations and isotopic ratios in river sediments. *Earth and Planet. Sci. Letters*, 4, 245-252.

Supervising Scientist (1984-85) *AAR Annual Report*, Aust. Govt., Canberra.

Tuan, Yi-Fu (1966) New Mexican Gullies: a critical review and some recent observations. *Annal. Assoc. Amer. Geog.*, 56, 573-597.

Williams, M.A.J. (1973) The efficacy of creep and slope-wash in tropical and temperate Australia. *Aust. Geog. Studies*, 11, 62-78.

Wolfenden, P.J. and Lewin, J. (1977) Distribution of metal pollutants in floodplain sediments. *Catena*, 4, 309-317.

Wolfenden, P.J. and Lewin, J. (1978) Distribution of metal pollutants in active stream sediments. *Catena*, 5, 67-78.

16

Complex Channel Response to Urbanisation in the Dumaresq Creek Drainage Basin, New South Wales

R.J. Neller

School of Australian Environmental Studies
Griffith University
Brisbane

I. INTRODUCTION

Wolman's model of stream channel response to urban development for the mid-Atlantic coastal region of the United States has strongly influenced geomorphic and planning thought on urban waterways for the past two decades (Wolman, 1967; Wolman and Schick,1967). He observed high suspended sediment loads and channel aggradation coincident with initial construction activities, and channel degradation during a later stabilization phase, when sediment loads declined to or below pre-urban levels. Since that time many studies have accumulated, and the importance of sediment migration on downstream channel morphology is also well established (Guy, 1974; Graf, 1975).

However while a change of channel reduction and enlargement in urban catchments is frequently recognised (Leopold, 1973; Guy, 1974; Graf, 1975; Robinson, 1976; Douglas, 1985), there is evidence of different channel responses. For example, Hollis and Luckett (1976) found that insignificant channel changes in the urbanising Canon's Brook catchment occurred between 1956 and 1970, despite a 2.2 increase in the mean monthly flood. This resulted from an increased sediment load. Warner and Pickup (1978) recognise both aggrading and degrading sections along the highly urbanised reach of the tidal Georges River, as does Hannam (1983) in a small Bathurst catchment. Gregory (1977) argues that the long-term capacity loss of Dumaresq Creek is due not only to an increased supply of sediment, but also to its character, and to the position of the urbanising area within the Dumaresq Creek drainage basin. Nanson and Young (1981) observed that small Illawarra coastal streams are reduced in size only if urbanisation is unaccompanied by more direct channel disturbances, and more often they are enlarged because of those disturbances. Neller (in press) found that gullying of small urban catchments in Armidale is triggered by channel disturbances such as channel reshaping. It is imperative that these examples be viewed not as special cases, but as part of an as yet undefined array of possible channel responses, which may amplify the simpler reduction/enlargement model.

Gregory (1977) recognises that much of the variability in channel response to urban development is a function of the position of the channel within the broader drainage network, since this factor can also reflect the influence of topography, proportion of paved surfaces and so on. Park (1977) has also drawn attention to the differing response of various sized channels to urbanisation. Many Australian cities are located in the lower reaches of much larger drainage basins and the total impervious area relative to this larger drainage basin is quite small. In contrast, the smaller tributaries of these can be largely urbanised. It is in these situations where tributary catchments are urbanised at different rates and at different times, thereby promoting complex channel responses along the mainstream, that Wolman's model needs further development.

The Dumaresq Creek catchment in northern New South Wales has been seen as an example of alternative channel response (Gregory, 1977; Neller, in press). The lower reaches of Dumaresq Creek flow through the city of Armidale, and although the total impervious area of this catchment is only about 3.0%, smaller tributary catchments affected by urbanisation can have greater than 20% impervious cover. This essay, after brief details on the catchment and methods, examines the complexity of channel response to urban development in this basin by contrasting the responses of its tributary catchments and that of the main channel prior to discussing a model of complex response.

II.   THE DUMARESQ CREEK DRAINAGE BASIN

The elongated Dumaresq Creek catchment on the New England Tablelands (basin relief 452 m; catchment area 120 km^2) is

part of the headwaters of the Macleay River. Rock types
include granites, basalt-capped lower divides and metamorphics which produce a wide range of sediment load.
Typical valley-side slopes in the mid to lower reaches are in
the order 0.01 to 0.07, whilst the slope of the channel bed
of Dumaresq Creek declines from 0.02 to 0.001. There has
been considerable land disturbance in this basin since the
commencement of land settlement in the early nineteenth
century, including clearing for pastoral activities, the
construction of an upstream reservoir, and urban growth in
the lower basin.

Dumaresq Creek is an ephemeral, low sinuosity, constrained alluvial stream. Storm runoff reflects the highly
variable rainfall pattern, with the five wettest days
frequently accounting for more than 30% of the total yearly
rainfall and more than 50% of the total yearly runoff.
During such periods Dumaresq Creek becomes a minor torrent
(Douglas, 1974), however little damage is done because the
floodplain is predominantly open space.

In the lower basin there is a marked contrast between
Dumaresq Creek and its much smaller valley-side tributaries.
The largest urban-affected tributary catchment, Martins
Gully, is 10.5 km^2 compared to 76 km^2 with Dumaresq Creek
at their confluence. Tributary channels cut through
colluvial deposits, angular and poorly sorted gravels with
a shallow sandy loam A_1, although alluvial fans and small ill-
defined floodplains are to be found along the mid and lower
slopes. They are susceptible to gullying, possibly because
of the high percentages of exchangeable sodium in the solodic
soils that occupy the lower slope positions (Imeson, 1978).
The slopes of these urban tributaries (0.06 to 0.07) are
considerably steeper than for the mainstream (0.002), but
decline sharply to 0.02 at the junction with the floodplain
of Dumaresq Creek.

III. METHODOLOGY

To assess the impact of urbanisation on these stream
channels a two-tiered approach was adopted; examination of
channel characteristics (channel geometry, boundary
sediments) of all stream courses in the lower Dumaresq Creek
valley, whether urbanised or not, and selection of five
sites (two small headwater catchments and three reaches of
Dumaresq Creek) for more detailed monitoring of streamflow
and channel processes (Table I, Fig. 1). Before 1978 there
was no permanent stream gauging station within the urban

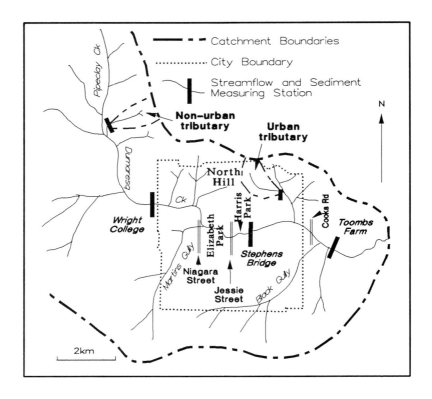

FIGURE 1. The lower Dumaresq Creek catchment.

area. Although occasional gaugings had been obtained
previously (Douglas, 1974), there had been no coordinated
monitoring of the tributary and mainstream flows. For this
reason additional records of flooding and of channel changes
were sought from newspaper records.

IV.   CHANNEL ADJUSTMENTS IN THE TRIBUTARY CATCHMENTS

On the valley-side slopes non-urban channels can be
classified as:

(1) broad, shallow depressions with ill-defined
boundaries, or

TABLE I. Experimental Design

	Tributaries		Dumaresq Creek	
	Location	Techniques	Location	Techniques
Streamflow	2 small catchments (<1km²) for 18 mths 1 urban 1 non-urban	Broad-crested compound rectangular weirs. Bristol pressure-bulb recorders. Ott current meter	Wright College - 64km² non-urban Toombs Farm - 102km² partly urban (both 18 months) Stephens Bridge - 78km² partly urban. 8 months	Natural channel control. Stilling wells and continuous chart recorders at Wright College and Toombs Farm; Bristol at Stephens Bridge. Ott current meter
Suspended Sediment Load	"	US U-59B rising-stage samplers suppl. with US DH-48 and UNE hand held samplers.* Vaccuum separation; 0.22µm millipore filters	"	US U-59C rising stage samplers suppl. with US DH-48 and UNE hand held samplers.* Vaccuum separation; 0.22µm millipore filters
Bed Load	"	Pit traps (40x20cm)	N/A	N/A
Channel Change	"	Erosion pins (measured every 10 days) and cross-section resurveys (every 4 months or after major runoff event)	At Wright College Toombs Farm and Elizabeth Park for 18 months	Cross-section resurveys (every 4 months or after major runoff event)

* Loughran (1976)

(2) stream channels, or

(3) gullies (usually discontinuous)

as previously recognised by Gregory (1977). All three
types of channel are interspersed along these small
·tributary profiles (Fig. 2). Channel morphology exhibits
considerable variability over short distances. The bed
profile consists of a series of knickpoints, or abrupt
steepenings of slope, often to 90°, and spaced on average
51 m apart. As these channels pass from steeper valley
sides onto the main floodplain there is an abrupt change in
their morphology, particularly for smaller channels.
Usually there is a reduction in channel size, and in some
cases the channel ceases to exist.

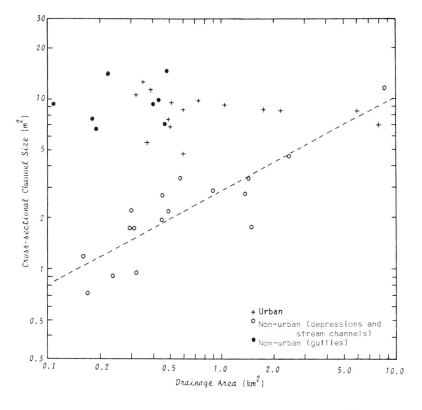

*FIGURE 2.   Cross-section channel sizes in tributary
catchments.*

In contrast urban tributary channels are consistently larger and less variable in size than the non-urban channels and depressions in catchments less than 5km^2. They are however of comparable size and shape to the non-urban gullied channels (Fig. 2). There is more variability in channel size amongst the urban channels as they cross the main floodplain. Some retain their dimensions, whilst others are similar to rural examples and are much smaller in size.

The degree of channel enlargement due to urbanization can be assessed quantitatively as the ratio of the present urban cross-sectional channel size to that which is estimated from non-urban stream channels for equivalent catchment areas (Hammer, 1972). This assumes that urban channels had similar geometry to non-urban channels prior to land-use change. This seems to be a reasonable assumption since non-urban gullies constitute less than 10% of the total non-urban channel length, and channels that were recognised as gullied prior to urban development (from 1943 aerial photographs) were excluded from this analysis. The 'enlargement ratio' for such urban channels averages 3.77 (ie almost a four-fold increase in channel size) with a clear tendency to decrease downstream (Table II). These data suggest a simple relationship between urban development and enlargement, that is more pronounced in smaller, more urbanised catchments. However, no significant relationship exists between the percentage of the impervious catchment (a surrogate measure of discharge), the proportionate change in impervious cover over the previous fifteen years, or the modified drainage density, with the enlargement ratio (r values all less than 0.25). Instead, there is a significant correlation of the enlargement ratio with the slope of the valley floor (r = 0.68), suggesting that once enlargement has been initiated, then morphometric rather than land-use attributes dictate the degree of response. This might suggest also that tributary valley slope may be close to threshold conditions as enlargement commences (Schumm, 1977).

Little is known about the process of enlargement and its initiation. It seems to involve the process of gullying because, not only are the geometric properties of enlarged urban channels and non-urban gullies the same, but gullying is currently occurring along numerous urban channels. Whilst concentrated storm runoff from impervious surfaces can induce gully erosion similar to that described above (Gregory and Park, 1976), it has been argued that direct channel disturbances (such as realignment and the building of road crossings) are probably more important factors

in these catchments (Neller, in press). This is based on
the widespread occurrence of enlarged channels throughout
the urban area irrespective of the degree of urban development, and on active gullying at sites of channel disruption
in both the urban and non-urban environments. This
apparently holds true for urban channels crossing the
Dumaresq Creek floodplain, as well as those draining the
valley-side slopes. The lowest reaches of Martins Gully
are deeply gullied upstream of the Niagara Street crossing
whereas the lower undisturbed reaches of Black Gully and
other small urban tributaries are much smaller in size.

*TABLE II: Enlargement Ratios for Urban Tributary
Channels*

Drainage Area (km^2)	Enlargement Ratio
0.32	6.31
0.35	7.19
0.37	3.13
0.39	6.34
0.49	3.96
0.50	3.54
0.51	4.92
0.60	2.31
0.60	4.22
0.74	4.49
1.77	2.47
2.38	2.00
6.00	1.11
7.90	0.78

Short-term changes in channel geometry continue to
occur along these gullied channels at sites where the
channel boundary is further disrupted, such as at sites of
infill and realignment, and at sites immediately downstream
of catchment disturbances, such as housing and road construction. A comparison of changes in the geometry of non-
urban and urban channels over an eighteen month period (Fig.
3) highlights the disruptive nature of these activities.
At site U1, unused road gravels were dumped in the channel
in February 1978, and subsequently removed during storm
runoff in March/April 1978. A change in the stormwater
drainage network caused further channel reduction in

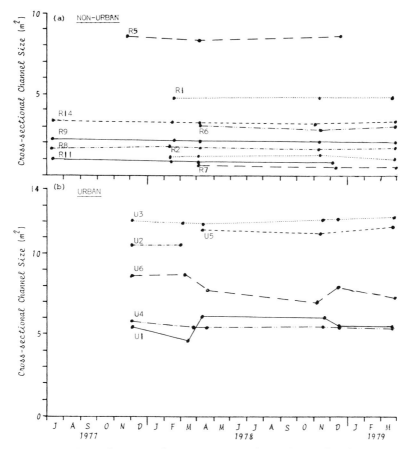

*FIGURE 3.   Changes in cross-section channel size -
tributary catchments.*

December 1978, when sediment-laden waters from a steep newly
graded, gravel road were directed into the channel at this
point. Site U6 was immediately downstream of a housing
construction site that was active in early 1978. Storm runoff
later in the year remobilized the sediments that had been
deposited during this period. At other sites, upstream
and downstream from these, and at all sites in the non-
urban environment, channel change was not apparent (Fig. 3).

   More precise measures of channel bank erosion and
knickpoint retreat were obtained using erosion pins and
an eighteen month period (Neller, 1988). The average
rate of bank erosion along the urban channel was 3.6 times
greater than that of the rural for the period (67 mm compared to 18 mm), whilst the rate of knickpoint retreat was
2.4 times greater (140 mm compared to 58 mm). However the

magnitude and incidence of storm runoff was much higher in
the urban catchment, and a comparison of similar sized runoff
events from these two catchments reveals no difference in
the rate of channel change.

Thus it appears that the small valley-side tributaries
of Dumaresq Creek are incipiently unstable, perhaps because
of the nature of the solodic soils along these waterways,
and that gully erosion is readily induced by direct channel
disturbance or local enhanced runoff. Once this gullying
has developed however, further disturbances to the channel
and catchment cause only short-term fluctuations in channel
sizes which otherwise appear to remain stable.

V.   CHANNEL ADJUSTMENTS ALONG DUMARESQ CREEK

Dumaresq Creek steadily increases in size below Dumaresq
Dam until it reaches the urban boundary. There it undergoes considerable reduction in capacity along its urban-
affected reaches (Fig. 4). This channel is on average only
63% of its expected size within the town boundary and only
45% of expected below town. At one site the channel size
is only 13% of that expected (Gregory, 1977). Reductions
are particularly noticeable immediately downstream of the
confluences with major urban tributaries. This loss of
capacity is due, not so much to decreased width, but to
reductions in depth, promoting an increased width/depth
ratio in the urban-affected reaches. The bed profile is
also affected. The average spacing of riffles at Wright
College is 80 m, or 4.7 times the channel width, whilst
it is much larger in town (103 m or 8.4 widths) and downstream (75 m or 7.1 widths). The distinction between pools
and riffles along the urban affected reaches is also
suppressed, with extensive sandbeds and minor braiding
apparent.

Downstream reductions in channel size have also been
observed by Nanson and Young (1981) in streams draining the
Illawarra Escarpment. In this basin, however, there are
no abrupt slope reductions along Dumaresq Creek or widening
of the valley floor that could naturally promote a smaller
channel. Downstream reduction in channel size is not a
regional characteristic of New England. Instead, Dumaresq
Creek has been subject to heavy siltation (McConnell et al,
1966; Burkhardt et al, 1968; Gregory, 1977). Survey records
for Stephens Bridge revealed that the 1975 channel was only
12% as large as that in 1927 (Gregory, 1977). A 1978

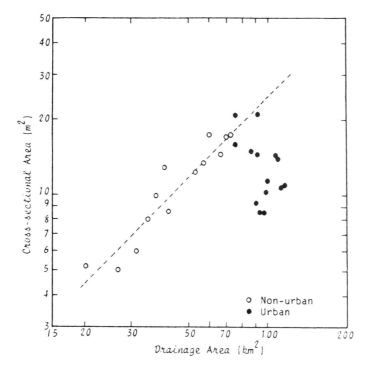

FIGURE 4. Bankfull cross-section areas of Dumaresq Creek.

resurvey of sites examined by Warner in 1967 (Burkhardt *et al*, 1968) gives rates of channel reduction for numerous reaches of Dumaresq Creek (Table III).   Abundant photographic evidence of siltation is also available.   The surveys and photographic evidence suggest that the rate of siltation is inconsistent with current channel dimensions and that the channel should be smaller.   However, the Armidale City Council regularly dredges Dumaresq Creek, and this probably accounts for the increased channel size at Harris Park in the centre and the higher rates of channel reduction at sites below town such as Cookes Road (Table III).

The sediment is urban in origin.   There is a rapid decline in channel bed material size at the upstream rural/ urban boundary.   Above the town the median particle size is around 40 mm, while at all urban sites, it is less than 10 mm.   This fining reflects a substantial increase in the proportion of sands and fine gravels.   Along the urban reaches there were large proportions of creamy chert, blue- metal and to a lesser extent, concrete fragments (at one

TABLE III.   Changes in Channel Size Along Dumaresq
Creek, 1967-1978.

Channel Reach	C/S Area at Bankfull (m^2)		% Change 1967-78	Rate of Change (m^2y^{-1})
	1967	1978		
Donnelly Street	20.74	15.93	-23.19	-0.44
Harris Park	18.80	22.03	+17.18	+0.29
Faulkner Street	19.26	12.41	-35.57	-0.62
Taylor Street	27.13	14.66	-45.96	-1.13
Douglas St. cut	40.60	31.58	-21.42	-0.78
de la Salle	16.02	8.56	-46.57	-0.68
Cookes Road	24.35	10.24	-57.95	-1.28
Average	23.78	16.49	-30.50	-0.66

1967 date from Burkhardt, Loughran and Warner (1968)

site the combined proportion of these three was 62.5%).
Creamy chert is used as fill for buildings and roads, but
it is mainly brought into Armidale from the adjacent Sumarez
Creek catchment.   It enters Dumaresq Creek via the stormwater
drainage network and the urban tributaries.   Construction
activities in tributary catchments are a notable source area
for these sediments (Douglas, 1974), as are unpaved roads
and driveways, roads that are neither kerbed nor guttered,
gardens, and the channel banks.   It has also been
demonstrated that the longer-term sediment yields within
the smaller catchments can remain significantly elevated
after urbanisation (Neller, 1985).

The siltation within the channel has increased the
incidence of overbank flows.   Newspaper records proved to
be an excellent source of such information (Table IV),
particularly since the late 1940s when housing encroached
onto the floodplain.   The recurrence interval during the
1970s was half of that expected from a regional flood
frequency analysis (Gregory, 1976).   The increased
incidence of main channel floods cannot be attributed to
the additional runoff from the urbanised tributary catchments, because that runoff would usually precede the main
floodwave generated upstream. The role of increased natural
runoff in the present flood-dominated regime also needs to
be considered in increasing the incidence of flooding (see

TABLE IV. *Incidence of Overbank Flooding Along Dumaresq Creek.*

Period	Number of Overbank Flows	Recurrence Interval (y)
1950-59	9	1.11
1960-69	10	1.00
1970-79	13	0.77

Erskine and Warner, this volume).

With the increased floodplain flows outside the diminished channel, there is reduced channel change along the urban-affected reaches of Elizabeth Park and Toombs Farm, in contrast to that of the non-urban reach at Wright College (Fig. 5). Here most change occurred during the period of maximum runoff (March/April 1978) with 40% of the 18 months of runoff recorded at this site occurring in three days, 30 March - 1 April. Peak flow was 13.7 m^3s^{-1}. Most change was associated with the removal or development of channel bars. Unfortunately Figure 5 does not fully reflect channel mobility at Wright College, because considerable reworking of the lower unvegetated channel boundary was also occurring. At Elizabeth Park and Toombs Farm there was minimum channel activity. The reduced size at EP5 resulted from a Council attempt to clear the heavily vegetated channel and to improve its capacity. At Toombs Farm change only occurred at a cattle crossing (TF6).

Thus Dumaresq Creek has undergone a metamorphosis from a larger, cobble/pebble, pool/riffle channel to the currently smaller, less mobile, partly sandbed channel, as a result of urban siltation.

VI. A MODEL OF COMPLEX CHANNEL RESPONSE

Quite clearly the model of urban channel reduction and enlargement proposed by Wolman (1967) bears little resemblance to the channel changes described within the Dumaresq Creek basin. It could be argued that this creek is still undergoing aggradation because of its urbanising tributary catchments and that once the process of urbanisation is complete channel scour will then predominate,

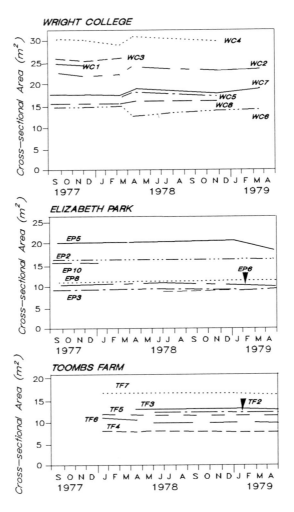

FIGURE 5.   Changes in cross-section areas along three reaches of Dumaresq Creek.

but for reasons given below this seems unlikely.   An alternative model, based on observations presented earlier and in this section, is proposed for this catchment (Fig. 6).

In the urban tributary catchments, channel enlargement is readily triggered by disruptions and proceeds without prior channel reduction.   The degree of enlargement is a function of the valley relief and perhaps valley-floor slope

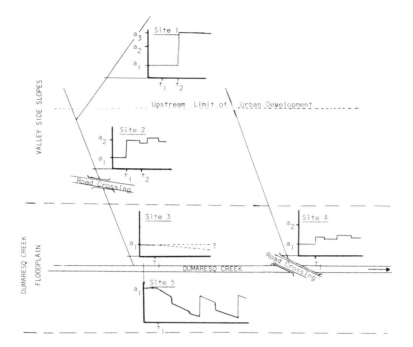

FIGURE 6. *Contrasting channel response to urban development in Dumaresq Creek drainage basin. X-axis is relative time; Y-axis is relative channel size.*

and since it is a gullying process, the enlargement at any site is abrupt. On the other hand, the stream migration of the gully head may continue for some years, and pass beyond the upstream limit of urban development (Fig. 6, sites 1 and 2). Since the development of newly-zoned urban lands can take place over many years, there are continual channel and catchment disturbances that cause minor short-term disruptions to the channel boundary (site 2).

As these urban tributaries cross the Dumaresq Creek floodplain they are often much smaller in size (site 3), and there are two possible explanations for this. Firstly, these dimensions may be retained from pre-urban days, in which case these channels could be viewed as transfer zones. Secondly, the downstream reduction in capacity may be enhanced by the downstream movement of urban sediments, and thus be viewed as a depositional zone. Due to a lack of data this issue remains unresolved. However, channel enlargement does occur along these reaches at sites subject to more direct channel disturbances (site 4).

To understand the process of siltation along Dumaresq
Creek (site 5) it is necessary to examine the tributary/
mainstream interactions in more detail. Channel slope is
obviously an important component of this because of the rapid
decrease in slope between the tributaries and Dumaresq
Creek. This loss in competence induces siltation, and this
is manifested in the reduced dimensions of Dumaresq Creek,
especially downstream of the confluence of major urban
tributaries.

An additional factor to these interactions must be the
magnitude and timing of the tributary inputs relative to
the mainstream flow. Because of the relative proportions
of impervious areas, the impact of urbanisation on Dumaresq
Creek is minor compared to that on its tributaries. Total
runoff between Wright College and Toombs Farm increases by
40%, but most of this can be attributed to the frequent minor
runoff from the urban tributaries when storm runoff is not
generated above the city. There were 3.7 times more runoff
events from the urban tributaries than from the non-urban,
constituting an additional 67 events in the eighteen months.
During the last three decades there have also been drought
periods of up to three years with negligible runoff upstream
of Wright College. Thus the bulk of urban runoff, and hence
the bulk of the sediment load, is discharged into Dumaresq
Creek when there is no additional runoff from upstream.
These sediment-charged waters lose their competence to
transport debris on the lower slopes of the mainstream, where
velocities are greatly reduced.

During the study period 332 tonnes of sediment were
transported downstream past Wright College, whilst the exit
load at Toombs Farm was only 228 tonnes. The urban
tributary contribution during this period was estimated to
be 1500 tonnes (based on some monitoring of Martins Gully,
Black Gully and some stormwater outlets to the creek plus
more rigorous observations from a small urban catchment).
This is equivalent to a reduction in channel depth of 40
mm per year, which is only about 50-70% of the rate of
channel reduction according to the siltation figures
presented earlier (Table III).

However, extensive deposition can also take place during
intense, localised, short duration cloudbursts over the
urban area. In March 1966, 58 mm of rain fell in 90 minutes
in the Martins Gully catchment, but only 15 mm fell in the
town centre only 2.1 km away. Runoff removed fences,
channel banks and road crossings in the lower Martins Gully
catchment. In November of that year an equally intense
thunderstorm over the same catchment resulted in 600 mm
sediment (mainly sand and pebbles, with some gravel) being

deposited along Dumaresq Creek 1.5 km downstream. Clearly
events of this type (there have been at least two others
in the past 30 years) have a significant, additional impact
on Dumaresq Creek to that of the more frequent, less
intense contributions. For this reason the siltation of
Dumaresq Creek is considered to take the form of a step
function (Fig. 6, site 5), with both on-going siltation
and occasional contributions from intense storm activity.

Short-term channel enlargement occurs in two ways: when
the creek is dredged by Council and by major floods with
low probabilities of occurrence. Unfortunately the only
record of the latter was in September 1969 when the long
abandoned and silted-up Jessie Street crossing was re-
exposed.

However, it is thought that Dumaresq Creek is likely
to remain much reduced in capacity for many decades to come.
The total urban proportion of the whole basin is still quite
low and there is much scope for further development.
Whilst the 1950s and 1960s were periods of rapid urban
growth, the 1970s was a period of relative stability.
Despite lower rates of construction activities during this
later period Dumaresq Creek continued to silt up (Table III).
This is probably because of the higher rates of soil loss
from the established low density urban areas compared to
that of the adjacent non-urban areas (Neller, 1985).
Moreover, the finer sediments and the higher nutrient inputs
from the urban area, the rapid colonization of channel bars
by vegetation and the increased transfer of flows via the
floodplain rather than the channel, all promote urban
channel stability.

VII. CONCLUSIONS

There has been little progress in recent years in
defining a general model of channel response to urban
development, and recent reviews continue to present a
"typical" channel response to a sequence of land-use changes,
with much less attention to the spatial variations throughout
a drainage basin of such a response. In this drainage
basin the position of the urbanising channels within the
broader drainage network, and the interactions between the
tributaries and the mainstream, are essential keys to
understanding the different response of these channels to
urban development. On the steeper valley-side slopes the
concentration of sodium in or near watercourses renders them
incipiently unstable and susceptible to rapid enlargement
following disturbances. The elongated nature of the Dumaresq

Creek catchment, the position of Armidale in its lower reaches, the nature and timing of sediment inputs from the urbanising tributaries and the relatively low rate and frequency of runoff from the dammed rural catchment ensure the long term reduction in channel size of Dumaresq Creek.

Both spatial and temporal variations in channel response to land-use change are shown in Figure 6, but it is quite specific to this drainage basin. Nevertheless, the approach could possibly be generalised to provide a more widely applicable model of urban channel response involving gullied headwater channels, zones of channel reduction and enlargement, and of sediment stores. The aim of such an exercise could provide urban planners with a more reliable model of the impacts of urban development on the fluvial system.

REFERENCES

Burkhardt, J., Loughran, R.J. and Warner, R.F. (1968) Some preliminary observations on streamflow and wash load discharge in Dumaresq Creek at Armidale, New South Wales, *Res. Ser. Applied Geog.*, No. 18, Univ. New England.

Douglas, I. (1974) The impact of urbanisation on river systems. *Proc. IGU Reg. Conf. and 8th N.Z. Geog. Conf.*, Palmerston North, 307-317.

Douglas, I. (1985) Urban sedimentology. *Prog. in Phys. Geog.*, 9, 255-280.

Graf, W.L. (1975) The impact of suburbanization on fluvial geomorphology. *Water Resources Res.*, 11, 690-692.

Gregory, K.J. ·(1976) The determination of river channel capacity. *Res. Ser. Applied Geog.*, No. 42, Univ. New England.

Gregory, K.J. (1977) Channel and network metamorphosis in northern New South Wales. In K.J. Gregory (ed.) *River Channel Changes*, John Wiley, Belfast, 389-410.

Gregory, K.J. and Park, C.C. (1976) The development of a Devon gully and man, *Geography*, 61, 77-82.

Guy, H.P. (1974) An overview of urban sedimentology, *Proc. Nat. Symp. Urban Rainfall and Runoff and Sediment Control*, July 1974, UKY BU106, Univ. of Kentucky, 149-159.

Hammer, T.R. (1972) Stream channel enlargement due to urbanisation, *Water Resources Res.*, 8, 1530-1540.

Hannam, I.D. (1983) Gully morphology in a Bathurst catchment. *J. Soil Con. NSW*, 39, 156-167.

Hollis, G.E. and Luckett, J.K. (1976) The response of natural river channels to urbanisation: two case studies from southeast England, *J. Hydrol.*, 30, 351-363.

Imeson, A.C. (1978) Slope deposits and sediment supply in New England drainage basin (Australia), *Catena*, 5, 109-130.

Leopold, L.B. (1973) River channel change with time: an example, *Geol. Soc. Amer. Bull.*, 84, 1845-1860.

Loughran, R.J. (1976) The calculation of suspended-sediment transport from concentration v. discharge curves. *Catena*, 3, 45-61.

McConnell, D.J., Montieth, N.H. and Berman, R.W. (1966) Soil conservation and changing land use in a small New England catchment, *J. Soil Cons. NSW*, 22, 29-41.

Nanson, G.C. and Young, R.W. (1981) Downstream reduction of rural channel size with contrasting urban effects in small coastal streams of southeastern Australia, *J. Hydrol.*, 52, 239-255.

Neller, R.J. (1985) Sediment sources in urban areas, *Search*, 16, 224-225.

Neller, R.J. (1988) A comparison of channel erosion in small urban and rural catchments, Armidale, New South Wales, *Earth Surface Processes and Landforms*, 13, 1-7.

Neller, R.J. (in press) Induced channel enlargement in small urban catchments, Armidale, New South Wales, *Environ. Geol. and Water Sci.*

Park, C.C. (1977) Man-induced changes in stream channel capacity. In K.J. Gregory (ed.) *River Channel Changes*, John Wiley, Belfast, 121-144.

Robinson, A.M. (1976) The effects of urbanisation on stream channel morphology. *Proc. Nat. Symp. on Urban Hydrol., Hydraulics and Sediment Control.*, Univ. of Kentucky, 115-127.

Schumm, S.A. (1977) *The Fluvial System*, Wiley, New York.

Warner, R.F. and Pickup, G. (1978) *Channel changes in the Georges River between 1959 and 1973/76 and their implications*, Bankstown Municipal Council.

Wolman, M.G. (1967) A cycle of sedimentation and erosion in urban river channels. *Geog. Annal.*, 49A, 385-395.

Wolman, M.G. and Schick, A.P. (1967) Effects of construction on fluvial sediment: urban and suburban areas of Maryland. *Water Resources Res.*, 3, 451-462.

17
Human Impacts on River Channels
in New South Wales and Victoria

Robin F. Warner

Department of Geography
University of Sydney
Sydney,   NSW

Juliet F. Bird

Environmental Science Centre
Melbourne College of Advanced Education
Carlton, Victoria

I.   INTRODUCTION

This essay is concerned with human impacts on river channels, whether deliberate, through schemes for water storage, drainage improvements and navigation enhancement, or inadvertent, associated with activities such as aggregate removal and changes in water or sediment loads. It contrasts with the three previous essays in this volume, which deal primarily with the hydrological and morphological consequences of intervention in the catchment.

The study of the human impacts in and near channels can be approached in two ways. The more traditional view has involved assessment of structures already in place, or of activities such as dredging, which are already completed. A second and more recent approach is to assess the probable impacts of proposed developments. This is now part of the environmental impact assessment procedure required by law. This may be termed prediction geomorphology, which uses experience of past impacts as well as present conceptual knowledge of geomorphological processes to assess future changes.

FLUVIAL GEOMORPHOLOGY OF AUSTRALIA
ISBN   0 12 735660 6

In this brief review of human impacts in channels, concern is mainly with the former approach - the impact of past endeavours - but examples of the latter approach demonstrate the geomorphologist's role in predicting the effects of development projects on associated river channels.

Impacts on the channel may involve changes to width, depth, width-depth ratio, channel capacity, roughness, slope, sinuosity, meander wavelength and so on. All influence the movement of water and sediment not only through the affected reach, but also up and downstream in ways that may disrupt the basic continuity of the channel, which in turn adjusts to the new situation. The purpose of this essay is to review the changes and their effects, with reference to NSW and Victorian examples. This is done in three main sections:

Deliberate modifications

    (A) channel regulation for water supply, power and navigation

    (B) channelisation and entrainment

    (C) vegetation management

    (D) urban drainage improvements.

Non-deliberate, or inadvertent modifications

    (A) aggregate extraction and mining disturbance

    (B) flood mitigation

    (C) additions to water discharge

    (D) loss of discharge

Impact assessment of planned developments.

II. DELIBERATE MODIFICATIONS

A. Regulation for Water Supply, Power and Navigation

This section is concerned with water-supply weirs, dams for water, power and other needs, and weirs, barrages and other structures for navigation. All structures are barriers to water and sediment movement. Impacts depend on their size relative to the modified river channel. While weirs may have only local impact on sub-bankfull flows, large dams may cause drastic changes not only in the valley floor directly inundated, but also in long sections of channel up and downstream.

1. Weirs. In southeast Australia, where rainfall-run-off regimes are highly erratic (see Finlayson and McMahon,

this volume), weirs have often been used to ensure water
supplies, particularly in the early stages of settlement,
eg. at Liverpool in 1823 (Fig. 1), to provide water for the
small township. The natural rock bar at Dights Fall, on
the lower Yarra, was reinforced to perform a similar role
for Melbourne, in the 1840s.

In NSW, on the Nepean River, it is possible to assess
the impact of many weirs built before 1911 over more than
100 km. These were to conserve riparian water for farmers
when flow was disrupted by dam construction in the headwaters. A detailed long profile of the river was surveyed
between Douglas Park and the Grose junction. A part
resurvey in the early 1980s (Warner, 1983; 1984) showed that
the weirs had generally acted as traps for fallen trees and
sand leading to a loss of channel capacity which was
counterbalanced by bank erosion in the immediate vicinity
of the weirs; flood frequency also increased. At Menangle
sediment deficiences downstream of the dams led to
scouring of the weir pond. This effect was particularly
marked in the Penrith weir pond which lies below the
Warragamba Dam.

2. *Dams.* Large dams affecting the whole valley floor
and long reaches of the channel above and below the area
inundated have been built for many purposes. Figures for
1976 (Walker, 1979) show almost 350 such structures on
Australian rivers. Large dams near Sydney, east of Perth
and in the Upper Yarra northeast of Melbourne provide water
for urban and industrial uses. Others, as in the Snowy
Mountains, serve for hydroelectric power generation, while
Keepit, Burrinjuck and Eppalock are irrigation dams. Many
are multiple purpose dams like Lake Eildon, north of
Melbourne, which is used for irrigation, hydro-electricity
and recreation. A few, like Glenbawn Dam on the upper
Hunter River, are primarily for flood mitigation and water
conservation (Erskine, 1985).

Dams greatly affect water and sediment discharge
patterns. Flow in the channel upstream is influenced by
the change in base level which rises to the height of the
spillway, affecting water movement far above the dam itself,
in the backwater reach. Coarse incoming sediments settle out
in the backwater or are deposited as a delta, while finer
materials may settle out as a layer on the former valley
floor. Deltaic sediments may be periodically incised at
times of low-water level, but the fines provide a valuable
continuous record of change in the catchment upstream, and
may be used to calculate overall denudation rates (Bishop,

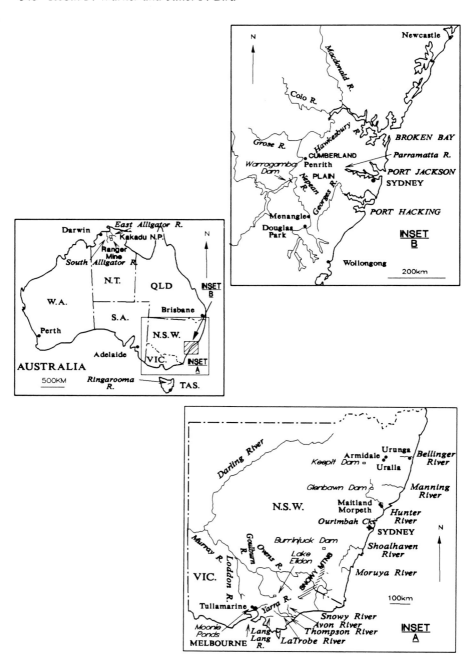

FIGURE 1.   Locations referred to in text.

1984) or the history and timing of individual sediment events, recorded in changes in sediment size, pollen or ash content (Clark, 1983; 1986).

Most attention has been focussed on reaches downstream from dams, where flows may be severely modified (see Petts, 1984, p.28 for summary of changes). In some cases total discharge is reduced, but a change in its distribution is more common. In the Murray-Darling system Dexter (1978) reported a decrease in winter-spring flooding, and an increase in summer-autumn flows as a consequence of regulation primarily for irrigation. Another common effect is a change in the pre-existing flood pattern, particularly a decrease in the size of low to moderate recurrence-interval floods (Erskine, 1985).

Sediment-deficient discharges over the dam may cause scouring if they have competence to move the sediment in the truncated channel. Such deepening is evident in the reach below the Warragamba dam referred to earlier. Erosion by the sediment-starved flows released from the dam has deepened gorge scour holes and increased the capacity of a 5 km reach above the Victoria Bridge, Penrith by over 660,000 m^3 in the last 80 years (Warner, 1983; Erskine and Warner, this volume). If flooding in the downstream reach is reduced following construction of the dam, bed armouring may develop as large material on the channel floor, formerly shifted in floods, becomes immobilised in the new, lower flood regime (Erskine, 1985).

Downstream of the scoured/armoured zone, channels adjust to the attenuated channel-forming flows by reducing their capacity (Petts, 1984). The effect may be marked where unregulated tributaries deliver sediment to the main channel (Erskine, 1985) but minimal in the absence of such supply. For instance, in the Hawkesbury, there is no reduction in channel size because sediment has been abstracted, some behind the Penrith weir but mostly by removal of sand and gravel for aggregate.

3. Weirs for Navigation. On some of the world's largest rivers weirs have helped to regulate water levels for navigation. River transport in Australia has generally been limited to short tidal reaches. The notable exception is the Murray-Darling system, where river steamers enjoyed a brief period of busy trading from the 1850s, when the discovery of gold, followed by agricultural development, brought a demand for goods to be transported to and from the inland. In the 1880s, they were replaced by rail transport (Williams, 1974). Both rivers have highly variable flow regimes, and low water levels commonly disrupted navigation, but it was not until the 1920s, long after the decline of the river trade, that weirs were constructed to try to improve the situation. Twenty five weirs were planned, but only 16 were ever built. Their economic value was limited to some recreation boating. They have more negative effects, inundating large areas of wetland and red-gum forest, and encouraging the rise of salty groundwater which feeds into the Murray, causing a deterioration in water quality (Walker, 1979).

B.   *Channelisation and Entrainment*

The term "channelisation" is used to describe the process of straightening and enlarging existing river channels in order to enable them to contain higher flows. "Entrainment" refers to the construction of new channels through swamps where defined flow alignments were generally absent. Both types of work are usually undertaken to try to improve flood control or land drainage.

1.   Channelisation.   Many rivers in lowland Victoria have been channelised and straightened by the cutting of by-passes across prominent meander bends. In the nineteenth century this was commonly done to eliminate flooding, but it was rarely successful and increasingly was perceived as of mixed benefit in rural areas. Later straightening was largely designed to reduce flood duration rather than frequency. On some rivers, such as the Latrobe, in Gippsland, flood control was attempted through flood cuts, channels excavated to half depth across meander necks. The intention was that they would only be occupied by flood flows remaining dry at other times. They were also supposed to reduce the risk of initiating erosion through channel steepening. Flood cutting was legally a simpler process, and reduced problems of farm fragmentation (Bird, 1979). Their stability depended on the nature of the alluvial soils; they only operated successfully where the sediments into which they were cut were sufficiently cohesive to resist permanent diversion of the river into the new, straighter course.

One effect of channel straightening is an increase in the slope of the river bed. This may cause the channel to cross a threshold of stability and convert from meandering to braided planform (Schumm and Khan, 1971). A more common consequence is the initiation of a rejuvenation head which migrates upstream, causing channel degradation and associated aggradation downstream.

It is doubtful if straightening alone leads to marked channel rejuvenation; straight channels are rare in nature, and the usual reaction of rivers to straightening is to restore the pre-existing gradient by enlarging remaining meanders or recreating the original ones. Strom (1962) described erosion along the Goulburn River, in northern Victoria, attributed to excessive straightening, but alternative explanations such as run-off changes due to modified catchment landuse were not considered. Most of the rivers which have been straightened in Australia are large with relatively low gradients and little relief available for incision. While some moderate bank erosion may be attributed to greater flow velocities associated with higher gradient, most of the more dramatic examples of channel incision are associated with entrainment works, rather than straightening alone.

2. Entrainment. In humid southeastern Victoria many smaller rivers flowed through swamps once inundated for several months each year. The absence of defined water-courses was probably due to the highly variable, but generally low discharge in smaller channels, and the abundance of sediment supply from the catchments. Prolific growth of dense swamp vegetation species such as swamp paper bark (*Melaleuca ericifolia*) may also have contributed by retarding local runoff. Recognition of the potential fertility of swamp soils led to attempts by nineteenth century farmers to reclaim them by constructing transverse drains in order to entrain the runoff and lower the water table. Small swamps were drained by individuals, but the larger wetlands like Kooweerup Swamp, southeast of Melbourne, were the subject of several large scale government schemes following the 1884 Land Act. Trunk drains were constructed for the main rivers which were supplemented by a rib-like pattern of tributary drains for smaller watercourses, and a network of parallel catch drains to receive local run-off (Key, 1968).

Artificial drains enlarged to accommodate the additional swamp flows caused rejuvenation of bed and banks in natural watercourses upstream, creating deep gorge-like cross profiles (Bird, 1980; 1985). The economic costs of incision have been considerable: bridges have been undermined, and productive valley-floor farmland destroyed or dessicated by falling water tables (Binnie, 1983). Downstream effects have also been severe. Sediment mobilised by erosion has

been redeposited downstream, obliterating the waterways
beneath bridges to cause increased flooding (Bird, 1982),
and in some cases spreading out to bury the once-productive
soils beneath a layer of sterile sand and gravel.

C.   *Vegetation Management*

A common early technique for "river improvement" was
desnagging, the removal of tree-trunks from channels. Snags
were very prevalent in many lowland rivers and they must
have had considerable influence on channel morphology. A
study is currently in progress at the River Murray
Labotatory in Adelaide to evaluate this (Lance Lloyd, pers
comm.). The prevalence of snags may be attributed to the
durability of riparian timber especially the river red gums
(*Eucalyptus camaldulensis*). The density of the timber is
another factor, making the trunks heavy and hard to shift
even in flood. Snags were so numerous in the Murray River
that they formed a major impediment to shipping. Isolated
snags were even more hazardous, especially those below the
water line. Extensive desnagging work was carried out along
the Murray Darling (Walker, 1979) in the 1850s and 1860s;
at the same time the supply diminished as the banks were
cleared of timber to fuel the steam boilers. Other rivers
in Victoria were desnagged to improve flow capacity,
particularly after the devastating 1934 floods, which
affected much of the southeast of the State.

Apart from the removal of dead timber from the channel,
many river banks were cleared of trees to reduce the channel
roughness and to improve flood flows. This practice caused
concern even in the very early days, and in 1803 the
Governor of New South Wales issued an edict designed to
prohibit clearing of riparian vegetation on the grounds that
its loss would increase the risk of bank erosion (Bolton,
1981). Such regulations were largely ignored, and the
destruction continued. Increasing velocities and loss of
the root network, important for ensuring bank stability
(Hickin, 1984), increased bank erosion. In the more extreme
cases, such as the River Avon, in Gippsland (SCRI, 1983),
it was enough to convert a small, meandering channel into
a wide, gravel-choked waterway in less than one hundred
years.

In an effort to control erosion the Victorian Government
encouraged farmers to replace the lost indigenous vegetation
with imported European willows (*Salix spp.*) in the hope
that they would rapidly establish and stabilise the channel

margins. Many environments proved highly favourable for
willow growth, but as they spread along the rivers it became
obvious that they might cause more problems than they solved,
particularly where they were able to establish mid-stream.
Strom (1962) estimated that up to 20% of the capacity of
the lower Snowy River was lost to willow growth. Not only
was the multi-trunked willow able to occupy much of the cross
sectional area of the channel, but it trapped sediment
and other debris leading to the formation of in-stream
islands or peninsulas, havens for vermin and noxious weeds.
After the 1934 floods the Victorial Government made an
emergency allocation of funds to support a further phase
of desnagging, this time to remove the willow trees so
enthusiastically planted 20 years earlier.

A much more flexible approach is adopted today in
management of vegetation in and adjacent to river channels
by state authorities. Desnagging, where possible, is confined to dragging tree trunks from the centre of the river
to the side, so that their biological role - fish habitat,
bird perching and preening sites, etc. - is maintained,
at the same time as using them to protect the banks. Existing
riparian vegetation is increasingly protected by fencing
to keep out stock.   Many sections of river bank have been
replanted, using native species such as water gum (*Tristania
laurina*) and crimson bottle brush *(Callistemon citrinus)*
where possible, but still resorting to willows in the most
severely eroded situations.

D.   Urban Drainage Improvements

Two major types of change may be distinguished; bank
stabilisation works are usually carried out to control local
bank erosion, while channel improvement works often involve
much more extensive remodelling of the slope and cross
section of the watercourse.

Bank stabilisation is often carried out in a piecemeal
fashion to protect locally critical sites where rapid erosion
threatens valuable real estate or facilities such as roads
and pipelines. It may involve sheetpiling, rock beaching,
placement of gabions, or concrete walling to protect the
footing or regrading of the channel margins, followed by
protection with stone facing or addition of groynes designed
to deflect the flow of the stream away from the eroding bank.
Much of the Cooks River, in Sydney, has been treated in this
way, as have parts of the Yarra in the eastern suburbs of
Melbourne. As a piecemeal process it is often cumulative;

protection of one section of the bank often leads to the development of erosion at another site nearby, and this in turn will require treatment.

Where the piecemeal approach proves inadequate, it may be necessary to resort to more drastic measures to restrict channel erosion. Numerous papers, such as those by Wolman (1967), Warner (1976), Whipple and Dilouie (1981) and Neller (this volume) have described the urban hydrological changes which lead to channel enlargement. As the proportion of impermeable surface, including concrete and tarmac, increases, infiltration rates diminish; at the same time insertion of street drains increases the rate of move- ment of water from catchment to channel. This increases total discharge and channel flow becomes more variable. Under low flow conditions the pollution levels tend to build up to an undesireable level, while the flood peaks cause rapid channel enlargement. Initially channel clearing and regrading may ameliorate the situation, but in intensively urbanised areas it is usually necessary to reshape the channel cross profile and line it with concrete. This diminishes boundary roughness, reducing the opportunities for pollutant build up, and enabling it to discharge flood- waters faster. An example of this treatment is Moonee Ponds Creek, which flows alongside the Tullamarine freeway, leading to Melbourne Airport. Ultimately, it may be con- sidered appropriate to divert the water into an underground pipeline, as in Elster Creek, in the southeastern suburbs of Melbourne. In this case the land over the top has, where possible, been retained as open space, where it serves as an emergency floodway as well as providing a recreational parkland at other times (MMBW, 1979).

III. INADVERTENT OR NON-DELIBERATE CHANGES

Where human activity has caused channel adjustments by indirect means the changes may be described as inadvertent. Examples are aggregate removal from the channel and flood- plains, flood mitigation works, and non-deliberate increases in water and sediment discharges.

A.   *Aggregate Extraction*

Extraction of sand and gravel from channels, for con- struction uses, creates discontinuities in bedload movement an effect which may be similar to that of dredging to create shipping channels. Increase of the channel cross-sectional area at the removal site leads to a decrease in water velocity and deposition in the holes created. If this accelerates sediment delivery from upstream, erosion can occur above the extraction site. Erosion may also be observed downstream, where a sediment deficit is created. This pattern has been described in the tidal Georges River (Warner and Pickup, 1978; Warner et al, 1977) and in the Hawkesbury Nepean (Warner, 1983; 1984).

Even when operations are off-river in alluvial storages some channel impacts may occur if care is not taken to maintain barriers between extraction site and channel. Warner and Pickup (1978) found that breaches in levees bordering extraction sites in tidal reaches of the Georges River at Chipping Norton and Liverpool led to increases in the tidal prism entering the upper estuary. The additional tidal movement resulted in velocity increases downstream, which caused bank erosion, particularly where channel dredging had increased underwater slopes. These ponds and their tidal water volumes created flushing problems which were revealed in dye-study simulations of sewage effluent from the Liverpool Sewage Works (Warner and Smith, 1979a; 1979b). In order to rehabilitate the site the State Government set up a statutory body, the Lake Chipping Norton Authority. The Penrith Lakes Development Corporation was created to perform a similar function on the Nepean River, but in this case extraction sites were non-tidal and not joined to the river. Problems here include: water stagnation, nutrient build up and other adverse water quality conditions. The Department of Water Resources has recently initiated a major study in New South Wales rivers to assess bedload movement and to ensure that the aggregate resources are not over exploited.

In the 19th century the search for alluvial gold caused widespread disturbance of river sediments in both Victoria and New South Wales. Initially the mining of these placer deposits was largely carried out by individual miners, and the effect was probably highly localised, but by the end of that century large companies were operating mechanical dredges which reworked enormous quantities of sediment both in the river bed and on the flood plain. No attempt was made to require the miners to reclaim the disturbed valley floor, which was left as irregular, hummocky terrain, with swampy hollows separated by barren ridges. There have been few studies in either New South Wales or Victoria on the impact of alluvial gold mining, though the effects may be

observed on many rivers, including the tributaries of the
Shoalhaven and Araluen Creek in New South Wales, and headwaters of the Ovens River at Beechworth in Victoria where
one of the original dredges remains in position. These early
mining operations were generally poorly documented, and it
is therefore difficult to quantify their impacts. However
in Tasmania, Knighton (1987) attempted to calculate the
extent of disturbance of the Ringarooma River by extrapolating from records of tin produced and knowledge of concentrations in the ore at various mines.

Mobilisation of sediment during mining greatly increased
sediment loads in many water courses. The form of
hydraulic mining that caused such devastation in rivers in
the Sierra Nevada, in California (Gilbert, 1917) was
fortunately practised very rarely in Australia, but there
are still a few examples of major changes due to mining.
Rocky River, near Uralla, in northern New South Wales, for
example, is now a sand-choked watercourse for many kilometres downstream from the mining site, while short sections
of the Loddon and the Yarra were permanently diverted
to obtain access to placer deposits. In 1905 the Victorian
Government realised the threat posed by alluvial mining,
and created the Sludge Abatement Board to deal with the
problem. Thereafter there was at least some attempt to
regulate the mining process and to enforce reclamation after
completion of mining.

A few rivers have also received increased sediment loads
from poor placement of mine overburden waste, particularly
coal. When the first brown coal pits were opened up in
Victoria's Latrobe Valley in the 1930s overburden was piled
at the edge of the Latrobe River. As the river eroded the
foot of the heaps it released large quantities of coarse
sand which can still be identified as a slug shifted a little
further down the river by each flood (Bird, 1979).

B. *Flood Mitigation*

The term in this context is used to refer to the construction of artifical levee banks and improvements along
natural levees to control the extent of inundation of the
floodplain. In New South Wales, following the disastrous
flooding in the early 1950s which marked the beginning
of a new flood-dominated regime, flood mitigation authorities
were set up in most of the coastal valleys. The major works
were levee building and strengthening to keep out low level,
frequent floods, and the construction of an improved network of flood drains in the flood basins, often fitted with

flood gates to prevent return flows. These evacuated excess
water more rapidly from fertile valley-floor farmland behind
the levees.

Retaining floodwater in the channel rather than allowing
it to spread across the floodplain concentrates energy,
thereby facilitating erosion of bed and banks. Effectively,
it means an increase in the magnitude of high frequency
floods, usually combined with increased channel depth and
decreased roughness, all conducive to higher flow velocities.
Amounts of erosion directly attributable to flood mitigation
works do not in general appear to have been great, probably
because floodplains are generally low energy environments,
and also because in many cases, such as the lower Bellinger
(Warner and Paterson, 1987) the levees only serve to contain
relatively low level floods. In the case of the Bellinger,
however, there is evidence that increased flood discharge
within the channel, partly attributed to flood mitigtion,
and partly to changes in climatic regimes since the mid 1940s
(Erskine and Warner, this volume) has caused gravel bars
to move downstream to within 11 km of the sea. This reach
was navigable by large boats only a century ago. In order
to minimise problems of enhanced channel flows following
levee construction, most modern schemes, such as that outlined by Gutteridge, Haskins and Davey for the Murray River
(GHD, 1987), suggest the siting of levee banks as far as
possible back from the river. This may protect farmland
or urban developments, at the same time as retaining parts
of the floodplain for storage of floodwater.

In some cases short levee banks are designed to
prevent floods entering chutes. Such chutes can divert the
main river if erosion is allowed to progress unchecked. An
example of a threatened diversion was the avulsion of the
Thompson River, in Gippsland, into a small anabranch, Rainbow
Creek, during floods in 1952. Had emergency remedial work
not been undertaken, it is likely that the Thompson would
have permanently diverted. Another example involved the
Lang Lang River, where a levee across a flood chute directed
additional floodwater down the main channel, contributing
to erosion which converted the formerly shallow, meandering
stream into a gulch 15 metres deep within a few decades
(Bird, 1980).

The threat of erosion has led to the inclusion of
designated floodways in some flood mitigation works. These
are broad, relatively shallow channels designed to
accommodate excess flows above a certain level. A floodway
of this type was constructed alongside the lower Hunter,
where it provides a shorter high-flow route for the river

between Morpeth and Maitland. Such channels are straighter
and therefore steeper than the natural ones they supplement.
When in use, water flows down them at higher velocity than
in the main river, and it may transport coarser debris,
injecting it back into the original river below the "normal"
end point, and creating a discontinuity in sediment size
which may destabilise the system and cause channel widening.

C. *Additions to Water Discharge*

Additions to water discharge may take the form of
controlled interbasin transfers, low level increments through
use of storm water drains to discharge water imported from
adjoining catchments for urban supplies, or of catastrophic
events like dam failure.

The longest established interbasin transfer scheme in
Australia is the Snowy Mountains Scheme, which involves
lifting water from the headwaters of the south-east flowing
Snowy River across the Divide into the headwaters of the
Murray. The impacts of the additional flows, equivalent
to 12% of the inflow into Lake Hume (Walker, Hillman and
Williams, 1978) on the recipient streams have been studied by
the NSW Water Resources Commission (1982). They found from
repeated cross-sectional surveys and study of air photographs that there had been a general increase in channel
size. Further down the Murray valley the redistribution
of water for irrigation has added to summer discharge, but
any effects on total discharge are likely to have been out-
weighed by the far more marked impact of rising water
tables, which have changed many rivers, described by early
explorers as occupying channels many metres deep, into
shallow, superficial streams (Bird, 1976).

Assessment of the impact of dam failure has to date been
only a prediction exercise in Australia, as no major dam
failures have occurred. A farm dam near Winchelsea in
Western Victoria breached after heavy rain in 1987, causing
large quantities of sediment from the earth dam wall to be
washed downstream. The initial gulch created by the
sudden discharge faded within a few metres as the sediment
was deposited to block the former outlet channel. In New
South Wales there has been concern for the safety of the
Warragamba Dam, where predictions of the Probable Maximum
Flood have recently been revised upwards. Evidence from
slackwater deposits below the dam (Riley *et al*, 1988 in
press) suggested that the highest flood was in fact about
5 m above the 1867 largest observed flood. Such a flood

would have a magnitude of about 40,000 m^3s^{-1}, compared to 18,000 in 1867. Assuming the dam itself did not fail this would involve an overbank discharge of 20,000 m^2s^{-1}, as well as very high channel velocities. However the recurrence interval of the PMF is reckoned to be between 1000 and 100,000 years (MWS & DB, 1985) so the risk is relatively low. Failure of the dam would mean the release of 2,000,000,000 m^2.

D. *Loss of Discharge*

Some data on the impact of the Snowy Mountains scheme on the Snowy River have been collected by the Rural Water Supply Commission of Victoria, and there is currently a proposal to prepare a full evaluation of the impact on flows and sediment discharge. Preliminary analysis suggests (Weinmann, 1987) that the effects of the diversions from the upper part of the catchment, which commenced in 1965, have been most pronounced on floods of moderate size, which have become much less frequent. Very low discharges are also much more common (Barry Jones, pers. comm.). The increased siltation occurring in the lower reaches of the river probably reflects aggradation in response to the diminished discharge, as well as increments to supply from erosion following catchment clearing.

IV. IMPACT ASSESSMENT OF PLANNED DEVELOPMENTS AFFECTING RIVERS

Prediction geomorphology is required in the preparation of environmental impact assessments for all developments which may affect rivers. In the past these were prompted by engineers with little regard for fluvial geomorphology. There have been notable exceptions, such as the late Professor C.H. Munro, the founder of water engineering in Australia, who began a tradition of mutual respect and co-operation between engineers and earth scientists in river engineering. Many instances can be cited, in which geomorphologists have been called upon to advise on engineering projects; their involvement in the hydroelectric and mining developments in Papua New Guinea is an example (Pickup, 1980; 1984; Pickup and Warner, 1984). They have also advised on potential rehabilitation of uranium mines in the Alligator Rivers area of Northern Territory (Pickup *et al*, 1987), where it was necessary to evaluate the long term fate of off-channel tailings storage, particularly since the area potentially affected lay in the Kakadu National Park, which had been nominated for World Heritage Listing. Other contributions by geomorphologists include work with the flood strategy plans for coastal rivers in New South Wales (Gutteridge, Haskins and Davey, 1979; 1980; 1981) and in the preparation of environmental guidelines for river management in Victoria (Loder and Bayley, 1988).

Contributions made by geomorphologists include assessment of flood magnitude and frequency changes, their impacts on channels and extent of flooding, and measurement of channel changes from maps, air photographs and other historical sources. They have also been involved in experimental work such as the tracer studies sponsored by the NSW State Pollution Control Commission to simulate sewage dispersal in the Georges River (Warner and Smith, 1979a; 1979b).

Geomorphological evidence has also been used to clarify definition of the 100-year flood level, as in the study of factory location problems at Ourimbah Creek (Sinclair Knight, 1983), which attempted to use the margin of the Pleistocene terrace as the flood boundary. In Victoria the 100-year flood is still adopted as the limit for many flood plain developments (Leigh, 1983), though its use has recently been abandoned in New South Wales. Geomorphologists have also been called upon to advise on impacts of gravel extraction in the Manning River (Sinclair Knight, 1987) and bank erosion problems in the Bellinger River (Cameron McNamara, 1984; Warner and Paterson, 1987). Much of this work is relatively inaccessible, since it is recorded in consultants reports to companies which may remain confidential documents. Studies which are more readily available because they are published in scientific journals include Pickup (1984), Pickup and Warner (1984), Warner and Paterson (1987) and Riley *et al*. (1988 in press). This is an expanding area of interest for numerous fluvial geomorphologists and one which will receive much more attention in future reviews of this kind.

V.   CONCLUSIONS

This essay has reviewed the direct human impacts on river channels in Australia, with particular reference to examples from New South Wales and Victoria. This bias in examples may be partly attributed to the authors' greater experience in these states, and partly to a real dearth of research into human impacts on channels in the less densely populated parts of the country. It is clear that the extent of impact in all parts of the country will increase in the future. A major cause will be the attempt to improve management of water resources in order to faciliate development in drier areas and to reduce the damage caused by periodic droughts in humid parts of the country. Rivers will also be increasingly affected by exploitation for other purposes including mining and recreation.

As technology advances the scale of this intervention will grow, so that the effects of individual developments will be felt through a large part of each river system. The range of variables encountered is likely to make it difficult for Government and planners to rely solely on techniques favoured by traditional engineering practices, notably the mathematical analysis of hydrological parameters, for prediction of the consequences of deliberate modification of rivers. Increasingly, the geomorphologists's skills at interpreting hydrological behaviour from existing landforms may be incorporated as a useful adjunct to numerical modelling. While the geomorphologists must continue to improve those interpreting skills in the light of past experience of comparable developments, they will also have to develop increasingly sophisticated techniques of analysis in order to enable them to predict the impacts of those modifications over the larger areas and longer time scales likely to be involved.

In many fields Australia has been able to import the results of overseas research as a background for work here, but in the case of river channel studies the development of a local understanding is critical for two reasons. One is the fundamental hydrological uniqueness of Australian rives, as outlined by Finlayson and McMahon. Of particular significance for evaluation of proposed modifications of rivers is the very high variability of flows encountered in most rivers here, which means that river channels may react differently to those in less variable environments. The other local characteristic is an historical one. Most Australian rivers are still adjusting to "un-natural" catchment conditions, a stage that was passed many years ago in other parts of the world, and they have therefore an inherent instability due to the indirect changes. The deliberate modifications outlined in this essay are being superimposed on a set of changes which in most cases are still in progress, rather than the established order of stability prevalent in many river systems in the old world.

While identification of reference catchments, including those as close as possible to "natural" (Macmillan, 1981) will provide a useful baseline from which to evalute the extent of channel changes, the work should be supported by on-going documentation of channel changes in modified environments to provide the data for an improved predictive geomorphology of river channels to be incorporated into major planned developments throughout the country.

REFERENCES

Binnie and Partners (1983) Master plan of works, Shire of Alberton River Improvement Trust.

Bird, J.F. (1976) Thomas Mitchell and the exploration of Australia Felix. *Proc. Roy. Geog. Soc. Aust.*, 77, 1-11.

Bird, J.F. (1979) Bank erosion on the Latrobe River and its lower tributaries. *Environ. Studies Prog. Publ. No. 267*, Ministry of Conservation, Vic., 69-97.

Bird, J.F. (1980) Some geomorphological implications of flood control measures on the Lang Lang River, Victoria. *Aust. Geog. Studies*, 18, 169-183.

Bird, J.F. (1982) Channel incision on Eaglehawk Creek, Gippsland, Victoria. *Proc. Roy. Soc. Vic.*, 94, 11-22.

Bird, J.F. (1985) Review of channel changes along creeks in the northern part of the Latrobe River basin, Gippsland, Victoria, Australia. *Zeit. f. Geomorph. Supp.*, 5, 97-111.

Bishop, P. (1984) Modern and ancient rates of erosion in central eastern NSW and their implications. In R.J. Loughran (ed.) *Drainage Basin Erosion and Sedimentation*, Univ. Newcastle and Soil Con. Serv. NSW, 35-42.

Bolton, G. (1981) *Spoils and Spoilers: the Australian Experience*, Allen and Unwin, Sydney.

Cameron McNamara (1984) *Bellinger River Morphological Study, Main Report*, Rept. for Bellingen Shire Council.

Clark, R.L. (1983) Pollen and charcoal evidence for the effect of aboriginal burning on the vegetation of Australia. *Archaeol. in Oceania*, 18, 32-37.

Clark, R.L. (1986) Pollen as a chronometer and sediment tracer, Burrinjuck Reservoir, Australia. *Hydrobiologia*, 143, 63-69.

Dexter, B.D. (1978) Silviculture of the river red gum forests of the central Murray flood plain. *Proc. Roy. Soc. Vic.*, 90, 175-191.

Erskine, W.D. (1985) Downstream geomorphic impacts of large dams: the case of Glenbawn Dam, NSW. *Applied Geog.*, 5, 195-210.

Gilbert, G.K. (1917) Hydraulic mining debris in the Sierra Nevada. *U.S.G.S. Prof. Paper*, No. 105.

Gutteridge, Haskins and Davey (1979) *Moruya River Valley, Flood Plain Management Study*, NSW Dept. Pub. Works.

Gutteridge, Haskins and Davey (1980) *Hawkesbury Valley, Flood Plain Management Study*, NSW Dept. Pub. Works.

Gutteridge, Haskins and Davey (1981) *Nambucca Valley, Flood Plain Management Study*, NSW Dept. Pub. Works.

Gutteridge, Haskins and Davey (1986) *Murray River Flood Plain Altas*, Vic. Rural Water Comm. and NSW Water Res. Comm.

Hickin, E.J. (1984) Vegetation and river channel dynamics. *Canadian Geog.*, 28, 111-126.

Key, L.M. (1968) Draining the swamp. In N. Gunson (ed.) *The Good Country: A History of Cranbourne Shire*, Cheshire, Melbourne, 136-153.

Knighton, A.D. (1987) Tin mining and sediment supply to the Ringarooma River, Tasmania, 1875-1979. *Aust. Geog. Studies*, 25, 83-97.

Leigh, C.H. (1983) Drainage and flood plain management in the Melbourne and metropolitan region. Paper at Inst. Aust. Geogrs. Conf., Univ. Melbourne.

Loder, J. and Bayley, I. (1988, in prep.) Environmental Guidelines for River Works. Rept. to Vic. Dept. Water Res.

M.M.B.W. (1979) The Development of Elster Creek Drainage System. Melb. and Metro. Board of Works, Rept. No. D-0022.

MacMillan, L.A. (1981) Indentification of pristine catchments in East Gippsland. Aust. Heritage Comm., Canberra.

M.W.S.&.D.B. (1985) Warragamba Dam Flood Protection Programme. Metropolitan Water, Sewerage and Drainage Board.

Petts, G.E. (1984) *Impounded Rivers: Perspectives for Ecological Management*, Wiley, Chichester.

Pickup, G. (1980) Hydrologic and sediment modelling studies in the environmental impact assessment of a major tropical dam project. *Earth Surf. Proc.*, 5, 61-75.

Pickup, G. (1984) Geomorphology of tropical rivers. I. Landforms, hydrology and sedimentation in the Fly and Lower Purari, Papua New Guinea. *Catena*, Supp. 5, 1-17.

Pickup, G. and Warner, R.F. (1984) Geomorphology of tropical rivers. II. Channel adjustment to sediment load and discharge in the Fly and Lower Purari, Papua New Guinea. *Catena*, Supp. 5, 18-41.

Pickup, G., Wasson, R.J., Warner, R.F., Tongway, D. and Clark, R.L. (1987) A feasibility study of geomorphic research for the long term management of uranium mill tailings. *CSIRO Div. Water Resour. Res.*, Rept 87/2.

Riley, S.J., Creelman, R., Warner, R.F., Greenwood-Smith, R. and Jackson, B.R. (1988, in press) The potential in fluvial geomorphology of a new mineral identification technology (QEM*SEM). *Hydrobiologia*.

SCRI (1983) The State of the Rivers. Standing Committee on River Improvement, Melbourne.

Schumm, S.A. and Khan, H.R. (1971) Experimental study of channel patterns. *Nature*, 233, 407-409.

Sinclair Knight and Partners (1983) *Definition of the 1 in 100 year flood plain of Ourimbah Creek*. Rept. for NSW Dept. Pub. Works.

Sinclair Knight and Partners (1987) *Proposed aggregate extraction, Manning River at Charity Creek, near Wingham*. E.I.S. for Alonbar Pastoral Co. Pty. Ltd.

Strom, H.G. (1962) *River Improvement and Drainage in Australia and New Zealand*. Vic. State Rivers and Water-Supply Comm.

Walker, K.F. (1979) Regulated streams in Australia: the Murray-Darling river system. In J.V. Ward and J.A. Standford (eds.) *The Ecology of Regulated Streams*, Plenum Press, New York, 143-163.

Walker, K.F., Hillman, T.J. and Williams, W.D. (1978) The effects of impoundment on rivers: an Australian case study. *Proc. Int. Assoc. Theoret. and Appl. Limnol.*, 20, 1695-1701.

Warner, R.F. (1976) Water and man in the city: modifications to hydrologic and geomorphic systems. *Geog. Bull.*, 8, 74-89.

Warner, R.F. (1983) Channel changes in the sandstone and shale reaches of the Nepean River, NSW. In R.W. Young and G.C. Nanson (eds.) *Aspects of Australian Sandstone Landscapes*, A.N.Z. Geom. Group Spec. Publ., No. 1, 106-119.

Warner, R.F. (1984) Man's impacts on Australian drainage systems. *Aust. Geog.*, 16, 133-141.

Warner, R.F. and Paterson, K.W. (1987) Bank erosion in the Bellinger Valley, NSW: definition and management. *Aust. Geog. Studies*, 25, 3-14.

Warner, R.F. and Pickup, G. (1978) *Channel changes in the Georges River between 1959 and 1973/76 and their implications.* Bankstown Municipal Council, NSW.

Warner, R.F. and Smith, D.I. (1979a) *The simulation of effluent movement in the tidal Georges River, NSW, Australia, using the dye tracer Rhodamine W.T.* Rept. for NSW State Pollution Control Commission, Bankstown Municipal Council, NSW.

Warner, R.F. and Smith, D.I. (1979b) The use of fluorometric tracers to study effluent movement in the Georges River, NSW, Australia. *Proc. 10th NZ Geog. Conf. and 49th ANZAAS Congress,* 48-53.

Warner, R.F., McLean, E. and Pickup, G. (1977) Changes in an urban water resource: an example from Sydney, Australia. *Earth Surf. Proc.,* 2, 29-38.

Water Resources Commission (1982) *Effects of the Snowy Mountain Scheme on the Upper Murray River.* Rept. NSW Wat. Res. Comm., Sydney.

Weinmann, P.E. (1987) *Effects of the Snowy Mountain Scheme on the lower Snowy River.* Discussion paper to Vic. Dept. of Water Res.

Whipple, W. and Dilouie, J.M. (1981) Coping with increased stream erosion in urbanising areas. *Water Resour. Res.,* 17, 1561-1546.

Williams, M. (1974) The lower reaches. In H.J. Frith and G. Sawer (eds.) *Man, Nature and a River System.* Angus and Robertson, Sydney, 140-159.

Wolman, M.G. (1967) A cycle of sedimentation and erosion in urban river channels. *Geog. Annal.,* 49A, 385-395.

Index

8 9 0 1 2 3 4 5 6 7
A B C D E F G H I J